2050: TheSingularHumanEra©

2050: The Singular Human Era

How to develop a future-ready mindset to distinguish yourself from the other 10 billion human beings (plus robots) by 2050

rafael IRIO

www.rafaelirio.com

To request permissions, contact the publisher at:**contact@rafaelirio.com**

Hardback ISBN: 978-0-578-38116-9

Paperback ISBN: 979-8-9858475-0-5

Ebook - Kindle ISBN: 979-8-9858475-2-9

Ebook - EPUB ISBN: 979-8-9858475-1-2

First Edition, March 2022

Book design by **Dani Groppo**
Final book review by **Danielle Siqueira**
and **Sarah Lavender**

www.rafaelirio.com

To all explorers and lifelong learners
around the world

Contents

Author's **Note**

A human being is a man, woman, or child of the species Homo sapiens, distinguished from other animals by superior mental development, power of articulate speech, and an upright stance. Homo sapiens, the first modern humans, evolved from their early hominid predecessors thousands of years ago. Their genetic code, physical traits, and superior cognitive and social abilities; for example, their ability to adapt to different circumstances, communicate, and collaborate on a large scale has offered them an advantage over other species.

Humans have gone through many revolutions that have fundamentally shaped our modern society and economies worldwide, pushing us to continue developing relevant skills to adapt to new scenarios. The global population will reach nearly ten billion people by 2050, increasing from 7.8 billion in 2020. As we expand our chances to live longer and healthier lives, we need to prepare ourselves mentally, physically, financially, and emotionally and remain flexible to adapt and thrive in an unpredictable future.

The exponential growth of technologies such as artificial intelligence (AI), robotics, Internet of Things (IoT), 3D-printing, virtual reality and augmented reality, biotechnology, and many others has substantially accelerated in recent years, and it will continue to do so. For example, six out of ten individuals worldwide expect that products and services using AI will profoundly change their daily lives in the next three to five years. Virtual reality and augmented reality headset shipments will grow to 32.8 million units by 2025, compared with 9.7 million units in 2021. As we look ahead, these technologies will not advance predictably or uniformly. In the next thirty years, the technological progress we will experience will be exponential, convergent, and more immersive than the progress made throughout the last 250 years, resulting in radical changes to how we live, work, learn, and connect with others. Technology will bring massive opportunities for us to explore and discover new possibilities continuously, but it will also bring challenges, anxieties, and unforeseen scenarios along the way.

The singular humans are forward-thinking individuals and lifelong learners. These two traits make them unique and distinguishable in the face of the rapid expansion of automation and robots. They embrace changes and turn unknowns into new possibilities to pursue their life goals. They do not feel overwhelmed by the explosion of data and media around them; instead, they make sense of information and translate knowledge to different contexts. Creativity is the new currency, and singular humans are the most prominent investors. They stack skills, knowledge, experiences and carry out the actions required to make their desired future a reality. The singular humans do not wait until amplified waves of change are about to crash down on them; instead, their curiosity allows them to anticipate change and strengthen their human skills to respond quickly. They build healthy relationships with digital technologies in ways to complement, optimize, and augment their human intelligence and abilities, including critical thinking and problem-solving.

We are entering a hyperconnected era where new technological revolutions will happen in months or even days as new technologies are being created and adopted faster than ever. Recycling your perspectives, skills, and behaviors in a continuous loop will enable you to respond to the ever-changing demands of a world that will be in constant flux. We are human beings in progress. Thus, developing a future-ready mindset allows us to bridge our present circumstances to the life we are hoping for.

As digital technologies become increasingly embedded in the fabric of our everyday lives and jobs, the distinction between reality and virtuality is blurring. Your mission is to preserve your identity, values, and purpose. Embrace the benefits of technology, but do not try to become or think like machines; this is surrendering your freedom and logical reasoning to them, especially as they continue to enhance their understanding of human emotions and behaviors. As machines, robots, and algorithms become more complex and sophisticated, hacking our biometrics and influencing our preferences, actions, and choices will become more valuable than targeting our computers or smartphones.

Today's children will endure more disruptions and transformations throughout their journey to adulthood than any generation has experienced previously. The prescribed path from education to work and then retirement will be dynamic rather than a straight line through life's milestones. Learning will need to flow a continuous cycle. As digital natives, the real challenge will not be embracing new technologies; instead, they will need to equip themselves with human-centric skills to survive an uncertain and turbulent future. Mastering collaboration, critical thinking, creativity, storytelling, and integrity while managing higher stress levels due to ongoing disruptions will be crucial to succeed in a changing world. Today's children hold the key to unlocking a better world. They will become the greatest voices against climate change and social injustices. They need to be encouraged to explore and discover new ideas and autonomously make decisions while still being held accountable for the outcomes.

2050: The Singular Human Era provides an overview of the critical waves of change that will continue to reshape the world over the coming decades. Understanding these changes is vital to anticipate and prepare ourselves to discover new possibilities without being paralyzed and anxious about an unknown future. Although our personal lives, jobs, and industries will be nearly unrecognizable by 2050, we hold the essential tool to navigate change over the coming years: our human ability to rise and adapt in the face of uncertainty.

How can we stay relevant in a world where AI will touch every job? What do we need to teach our children so they can thrive in an uncertain future? How do we prepare for a longer life as life expectancy increases? How can technology unite the world instead of amplifying the disparities between the rich and poor? These are some of the questions I aim to answer throughout this book. The collection of topics I present here serves as the basis for continued exploration and further discussions around new technologies and personal growth. As singular humans, we are always in progress; my role is to provide meaningful, compelling, and trustworthy content to contribute to your lifelong learning journey. I hope the information provided throughout the book serves as a foundation for you to build the bridge between where you are today and the life you want to live in the future. Don't wait until the next wave of change is ready to hit you; you must anticipate, prepare, and adapt to new scenarios and learn from your successes and failures so you can respond to the ever-changing world. I ask you to avoid becoming distracted by short-term irrelevancies and misinformation. By noticing where the world is heading and making sense of the rapid technological advances, you can develop the necessary skills to thrive in a fast-changing job market. I believe technology will continue to be a key avenue for humanity's growth throughout the twenty-first century; equally important, I believe the focus moving forward should not be on optimizing profits alone but on prioritizing human optimization. I hope you find value in the key topics discussed here and let them inspire you to build a more fulfilling and purposeful life. Above all, I want you to remain optimistic and motivated about the future and its new and unknown possibilities.

Get busy with your future!

Rafael Irio

Preface
A Day in 2050

Work bears little resemblance to what it was at the beginning of the twenty-first century. Most jobs have disappeared due to the rapid advances in AI and automation. These technologies have drastically transformed industries and made millions of workers redundant. In contrast, there has been a tsunami of new jobs created to manage new technologies. As a result, companies now find themselves in a war to attract professionals. The open positions are for the singular humans who possess specific human skills such as creativity, communications, critical thinking, integrity, adaptability, and persistence, instead of technologically or scientifically skilled.

Earlier this century, the focus was on building hybrid work models; in 2050, virtual realities and immersive technologies have made this possible. Today we have holograms and avatars representing us anywhere. However, humans have realized that to live purposefully, freely, and happily, the key is not where you work but what you do for work. As a result, we see many courageous explorers and forward-thinking individuals imagining and building the future in which they want to live, chasing experiences and the unknown over normalcy.

The abundance of data has made some people more intelligent, while others have been unable to synthesize the explosion of information around them. Everything that we do leaves a digital footprint. Smartphones are now redundant and hard to find—many people opt for a brain-computer interface, meaning no touchpad, no screens, and no keyboard, connecting minds and thoughts to the internet and the cloud. As digital technologies blur the lines between the physical, digital, and biological worlds, some people have search engines embedded in their brains. Still, interestingly, their top research is "how to become human again." More people enjoy the benefits of the digital world, but the divide persists; while some people are buying intelligence and health, vacationing on Mars, and owning the world's data, others are still learning how to write and send an email.

Home is now a command center, and one could easily believe it's a reality show as we are constantly under surveillance. It's filled with IoT sensors, smart furniture and appliances, and virtual assistants powered by AI that sound and behave like humans. All these devices are converged and directly connected to the internet. They can recognize our feelings, nutrient levels, and body temperature and create pleasurable moments, like adjusting air-conditioning levels, playing songs to boost our moods, or even dictating diets based on our health conditions.

Telemedicine in 2050 allows us to access any doctor globally, virtually anytime we want. Moreover, AI, algorithms, and wearable devices monitor our biometrics and detect diseases more precisely than human doctors. Nevertheless, doctors and many other human professionals remain vital since machines cannot replicate human empathy, compassion, trust, and sensitivity.

Our cities are hyperconnected and more digitally integrated. However, humans are not the only inhabitants. Machines and robots are integral parts of the ecosystems in most megacities. The number of sensors, devices, and drones tracking and monitoring traffic, logistics, security, health emergencies, and air quality is larger than the number of residents living in these cities. Physical forms of identification like driver's licenses are nonexistent—facial recognition systems are responsible for personal identification and security surveillance. People no longer own cars and transportation has become environmentally friendly as cars and flying drones are fully electric and autonomous. We see a substantial reduction in traffic congestion, accidents, and pollution in many cities worldwide. Green city infrastructure, from parks, tree-lined streets, to green roofs, is a priority, just like security, education, and healthcare. Large-scale vertical farming has become a vital source of food production since it uses a fraction of the land and water to feed a large portion of the population. Renewable energy, led by solar and wind, is responsible for most cities' electricity sources.

Education is online, mobile, and on-demand. We can now learn anywhere, anytime. AI systems can identify our needs and provide highly personalized content and learning. Virtual reality glasses have replaced physical classrooms, allowing students to teleport their eyes and ears to different places without leaving their homes. They visit and explore the world at their convenience. Physical schools and colleges still exist, but AI, cloud computing, virtual reality, big data analytics have transformed them into hubs for cutting-edge research and collaboration. Instead of requiring students to spend hours in listening mode, physical locations for education enable students to collaborate, apply, and share knowledge.

In previous centuries, people followed standardized education and fixed timelines. Today, lifelong learning is a requirement for survival in the workplace. Learning is intentional, and the winners in 2050 approach every experience as an opportunity to grow. There is no shortage of new experiences, challenges, or possibilities as breakthrough innovations are developed and adopted in record time. Humans must acquire new abilities faster and more frequently to keep up with the amplified technological disruptions and changes in their personal and professional lives.

The best ideas are those intended for everyone. The ideas fostering sustainability, unity, and solidarity succeed and create a prosperous world. The positive news is that expressions like "Thank you!", "How are you?", "I'm sorry!" and "I understand you!" are being used between people more than before. Amid the frequent waves of change, flexibility, empathy, sociability, humility, and integrity have remained crucial for adapting to new scenarios.

We must constantly remind ourselves **the future is all about people.** Technology will continue to drive humanity's progress, but **human's innovative mindset, fresh perspectives, trustworthiness, values, and curiosity** will continue to be unparalleled.

1 From One Billion to Ten Billion Human Beings

In the early 1800s, one billion human beings inhabited this planet. According to some projections, the global population will reach nearly 10 billion people by 2050, from 7.8 billion in 2020. New technologies, social, economic, and environmental transformations will reshape the world in the next thirty years. To flourish in such an unknowable future, we must prepare ourselves to embrace ever-changing scenarios; changes will keep coming like waves in the sea. Thriving in this hectic world will require anticipation, preparation, flexibility, and courage to adapt to future revolutions in our lives, jobs, and education.

The term "revolution" means a sudden, radical, or complete change—an activity or movement designed to effect fundamental changes in the socioeconomic situation. Since 1760 the world has experienced major industrial revolutions that have significantly shaped humanity, transforming largely rural and agricultural societies into industrialized and urban ones. Each revolution has introduced challenges, opportunities and created new technologies that have changed civilization and the way humans live and work. Historically, people shifted to healthier diets, leveraged improved transportation networks, more efficient communication methods, and accessed better housing and more affordable goods.

The invention of the steam engine during the First Industrial Revolution (1760–1850) is perhaps one of the most significant events during that time. Steam engines to power manufacturing processes yielded more productive and efficient factories. The Second Industrial Revolution (1850–1970) was characterized by rapid industrialization, standardization, and mass production. In addition, the expansion of railroads, large-scale iron and steel production accelerated during that time. In the 1970s, the world transitioned to the Third Industrial Revolution, also referred to as the Digital Revolution, characterized by the spread of automation, electronics, semiconductors, personal computers, and the internet.

Today we are living in the Fourth Industrial Revolution, defined by Klaus Schwab, founder and executive chairman of the World Economic Forum, as "the fusion of technologies that blur the lines between the physical, digital, and biological spheres." As curious and courageous human beings, we need to be prepared to respond quickly to the opportunities and adversities that will emerge from the digital transformation. Schwab believes "the changes are so profound that, from the perspective of human history, there has never been a time of greater promise or potential peril." This newest industrial revolution is significantly altering every aspect of our lives, everywhere in the world. While digital technologies in recent decades were already growing accelerated, many digital transformations strategized for years ahead were forced to emerge in the crisis state caused by the COVID-19 pandemic. As stated in the Global Trends 2040 report published by the National Intelligence Council, "the ongoing COVID-19 pandemic marks the most significant, singular global disruption since World War II, with health, economic, political, and security implications that will ripple for years to come." The pandemic has tested the resilience of individuals, communities, companies, and governments. It exposed the fragility and inequalities in our financial, healthcare, and social systems.

During the previous industrial revolutions, new technologies and inventions have led to significant transformations of social, cultural, and economic structures. However, looking forward, emerging technologies will not advance predictably or uniformly. In the next thirty years, we will experience more change, driven mainly by rapid and disruptive technological advances, than humanity has experienced over the last 250 years. It took millennia for society to transition from the agricultural to the industrial era and centuries to progress from the industrial to the information age. We are shifting now from the information age to a hyperconnected period where new technological revolutions will occur in months or even days.

As Earth becomes a more crowded planet, we must constantly remind ourselves the future is all about people. Technology will continue to drive humanity's progress, but human's innovative mindset, fresh perspectives, trustworthiness, values, and curiosity will continue to be unparalleled. As change is the only constant and technologies such as the internet, social media platforms, virtual reality, and augmented reality allow us to break geographical limitations, there will be no universal way to handle challenges. To adapt to an ever-changing and diversified world, you will need to think globally and act locally—being aware of social and historical biases and welcoming new cultures. As inequalities, discrimination, and the digital divide increase in many parts of the world, cultivating social awareness and a better understanding of cultural differences will aid you in identifying how you contribute to your community.

As the world embraces digital technologies, the barriers between countries, economies, and people will shrink even further. Respecting different perspectives, becoming an active listener, and empathizing with others—including those from diverse backgrounds and cultures— will be the foundation to build a more equal and ethical future.

Discontinuity will define the next thirty years. New technological, social, financial, and cultural revolutions will emerge faster and more frequently. These revolutions will create cascading global challenges that will repeatedly test our resilience and adaptability.

The Next Era **of the Internet**

It is difficult to imagine the world as we know it today without the internet. Since its creation, it has enriched our quality of life, removed the boundaries between societies and economies across the globe, allowed people to consume and share information in real-time, and acquire knowledge anytime and anywhere. In addition, the internet has helped consumers access more products and services cheaper and faster while also enabling them to become more aware of environmental, social, and governance strategies adopted by corporations. At the beginning of 2021, there were 4.7 billion active internet users worldwide—nearly 60 percent of the global population—up from around 50 percent in 2018. Moreover, mobile devices have made access easier nowadays; 93 percent of users access the internet that way.

We are in the midst of a significant digital transformation that will radically redefine how we connect ourselves to the internet. Not long ago, we used our screens (desktop computers) to access the virtual world. Today we use our hands (smartphones) or voices (virtual assistant devices) to stay online. Soon we will use our faces and eyes to immerse ourselves in the digital world using virtual reality and augmented reality glasses or contact lenses. Looking further into the future, we may become symbiotic with machines by linking our brains to the internet in real-time, using technology such as brain-computer interfaces, which would allow humans to control digital technologies using only their minds.

Inventor and futurist Ray Kurzweil has predicted that our brains will seamlessly connect to the cloud by 2035. New technologies ranging from noninvasive devices such as headsets and glasses to implanted brain sensors and chips will enable more seamless integration between humans and the digital world. Companies are working to transform what seems to be science fiction into reality and reinvent new forms of how people will communicate with machines in the future. For example, Neuralink, a neurotechnology company founded by Elon Musk, is researching and developing high-bandwidth brain implants that can communicate with technology. People with neural conditions and disorders could use the technology to regain independence through the control of computers and mobile devices. Still, Musk has also predicted it could enable human "symbiosis with artificial intelligence."

As we navigate the hyperconnected era, characterized by the widespread use of devices connected to the internet, 5G (or the fifth generation of cellular network technology) serves as the backbone for a more integrated world. It provides much faster and more stable wireless networking and connects more devices than previous generations. 5G can support up to one million devices per square kilometer, a far larger coverage span than the mere sixty thousand devices possible with 4G networks. Faster and more reliable networks are essential for enabling access and safe operation of other breakthrough technologies such as IoT, virtual reality, augmented reality, gaming, and autonomous vehicles. However, the race for the 5G successor is heating up. Although 6G is at least a decade away from launching, it holds the potential to revolutionize connectivity and communications. According to an article published on Bloomberg.com, the technology could enable "real-time holograms to flying taxis and Internet-connected human bodies and brains [. . .] not only can people communicate with things like furniture and clothes, but those gadgets can also communicate among themselves." As Ari Pouttu, a professor at the University of Oulu in Finland, states, "Let's throw the smartphones away and think about how humans interact with infrastructure [. . .] You can have smart surfaces with electronics in them, smart glasses so you can have devices everywhere."

The United Nations aims to achieve universal internet access, defined as 90 percent of the global population, by 2050. As online users increase exponentially in the coming decades, internet access may become an undisputed universal human right. Ensuring universal internet access in the next three decades is critical. Slow, unreliable, or nonexistent online access restricts the economic growth of rural and remote areas. Moreover, it increases the digital divide between developed and underserved communities. People cannot access real-time information or education to expand their skills, so their opportunities to access higher-paid jobs and quality education are limited.

Connecting the current offline population will be an arduous undertaking. Many of the remaining unserved areas are geographically challenging to reach due to rough terrain or remote location. Thus, raising providers' costs and pushing broadband services further out of reach of low-income households. However, more affordable and accessible internet is becoming a reality in parts of the world, thanks to satellite technologies emerging as an alternative to expanding access at lower costs to remote locations across the planet. For example, Starlink, a satellite internet constellation operated by SpaceX, has launched 1,735 satellites into orbit since 2019. According to the company: "Starlink is ideally suited for areas of the globe where connectivity has typically been a challenge. Unbounded by traditional ground infrastructure, Starlink can deliver high-speed broadband Internet to locations where access has been unreliable or completely unavailable."

Virtual reality and augmented reality are quickly taking off. Soon these technologies will redefine the way we access the internet, consume information, buy products, interact with others, and educate ourselves. These technologies will connect the physical world with the digital universe, making connectivity more immersive and interactive. In the coming years, instead of subscribing to a newspaper or buying a physical book, you will be able to immerse yourself in the story, moving away from reading to being present—experiencing and sensing any given scenario. Virtual reality will allow you to visit distant countries or planets, experience concerts from the front row in real-time from anywhere, or put yourself in the driver's seat of a race car. Search engine bars will disappear as you shift from chatting with virtual assistants to viewing and traveling virtually to any location. Soon we will move to a screenless future. Instead of using our smartphone or computer to search, we will be able to use our eyes (or smart glasses or augmented lenses) to find anything visually—products, locations, people, receipts—without staring at a screen or becoming distracted by a smartphone. Searching will move from typing and reading to seeing and thinking. Social media will no longer involve watching, clicking, or liking; its uses will include presence and experience. Virtual meetups with friends and family can happen anytime, from anywhere in the world.

Instead of scrolling through feeds, videos, or photographs on a two-dimensional screen, you will be able to interact with other people and avatars in the virtual world. You are likely to have more avatar friends than actual human connections. Most of them will represent companies and products. Some will be very nice and helpful and create a valuable digital experience for you, but other avatars will have only one goal: influence you at any cost. Looking at online advertisements and scrolling through websites to find products or services will give way to real-time and tailored experiences in the virtual universe. You will be able to engage and interact with the brands you trust, navigate a virtual shopping center, and test products with the option to personalize them.

As the internet becomes intertwined with our lives and digital technologies merge reality with virtuality, our attention will increasingly shift from the physical to the digital world. Going forward, we will need to routinely ask ourselves how our virtual life, experiences, engagements are benefiting our real-life. Every connected device and sensor in our home or body will be a potential access point to our private and personal data. Once digital technologies, fueled by AI, machine learning, and big data analytics become more sophisticated in reading and understanding us, it will be far easier for machines and algorithms to monitor and anticipate human emotions, attitudes, and decisions. Today's increased concerns about cybersecurity and malicious activities on the internet, such as hacking our smartphones or computers, will become tomorrow's concerns over hacking our biometrics and cognitive data. As you immerse yourself in the digital universe, your sensitive and personal data can be widely tracked, stored, and made available to companies and governments. Knowing your preferences, who you meet, what diet you follow, your daily habits, vulnerabilities, and what makes you happy, sad, or anxious can be more attractive and profitable than your address, passwords, emails, and credit card information.

As intelligent machines and algorithms monitor our lives increasingly, every interaction, comment, and step leaves a digital footprint. These insights empower companies and algorithms to create personalized campaigns to flood us with highly targeted content and advertising for products and services. As the race for our attention accelerates, many companies will try to influence our interests and decisions through digital surveillance, manipulation, and persuasion, making us their most valuable products rather than loyal consumers. Some governments, especially authoritarian regimes, can use new technologies and social platforms to favor themselves and their interests by disseminating misinformation, political polarization, and voter persuasion.

As AI and machine learning understand more about our lives and preferences, it is imperative to stay aware of who you are, what distinguishes you as a human being, who you want to become, and the life you want to build. Increasing your self-awareness by recognizing your motivations and what you stand for will be critical to avoiding the increased persuasion, manipulation, and polarization strategies that could harm your most vital human assets: identity, autonomy, values, and purpose.

Knowing how to distinguish reality from virtuality and what is pertinent versus unimportant will be vital so you can think independently despite the increasing presence and influence of AI, algorithms, and social media outlets. In addition, self-control—the ability to regulate your emotions, thoughts, and behaviors in the face of temptations and impulses—is a core skill to navigate the next era of the internet. To embrace and enjoy the digital world safely, you will need to question whether your online activity is beneficial or detrimental to your well-being and those around you. Finally, ensure the time you spend in the virtual world aligns with your long-term goals and intentions.

The internet will continue to power economic growth and innovation over the following decades. As more users navigate the online world, privacy and data protection must be enhanced, including how AI is developed and used by companies, governments, and individuals. Policies to govern the development of intelligent algorithms are necessary to prevent loss of trust, unfairness, and unethical behaviors in the ever-changing and growing cyberspace. As internet users entrust more of their sensitive and personal data to companies, they will need more transparent and rigid data privacy and security laws to protect it. Governments, legislators, and technology leaders are moving ahead with regulatory frameworks to protect the public. One example is the creation of the General Data Protection Regulation (GDPR), a set of rules designed to give European Union citizens more control over their data. Joining this global trend, Brazil recently enacted its omnibus law to govern the use of personal data, called the General

Data Protection Law. Similar to GDPR, the rule applies to any business or organization that processes people's data in Brazil, regardless of where that business or organization itself might be located. In the United States, the California Consumer Privacy Act requires companies to disclose what personal information they have on file and what they do with it. In addition, it offers consumers the option to ask businesses to delete their data. The European Commission has proposed new regulations and actions around AI systems. The proposal aims to ensure these systems are developed and used fairly and ethically. For example, the European Commission's AI regulatory proposal classifies AI systems under three distinct categories of risks: *unacceptable,* which includes manipulative or exploitative techniques that cause harm, specifically those used in public spaces for law enforcement; *high,* used to evaluate customer creditworthiness; and *limited,* including chatbots, computer games, and spam filters.

The internet and AI present a pivotal opportunity to drive human progress. Ensuring policies and regulations are responsive as digital technology rapidly advances is necessary to protect citizens' data and societal well-being. While also allowing companies to continue driving breakthrough innovations, remaining competitive, and pivoting their business models without slowing the research and development of new products due to outdated compliance protocols.

Data will be generated from everywhere. Cars, smartphones, bodies, minds, homes, and cities will be sources of massive amounts of information that will grow exponentially and flow at an unprecedented speed over the internet. In the years ahead, the expression "less is more" will have a greater level of importance as to how you synthesize the ocean of insights coming your way will aid you to understand the world around you. The amount of information and media surrounding us will be impracticable to process and retrieve; you can leave that for the machines, which can process data faster and more accurately than we can. Your job will be to focus on what matters most to you. The key to an easy digital future is not so much accessing a tsunami of data but how you comprehend and translate it into new contexts, scenarios, and ideas.

In the era of information, misinformation, and data explosion, becoming a critical thinker will help you endure the overwhelming amount of information you see, read, and hear daily. The key is to focus on relevant, trustworthy, and meaningful content by evaluating data and identifying possible connections to draw logical conclusions, arguments, and shape ideas. Applying rational analysis of facts and evidence to form your judgments, free of external persuasion, and being aware of your biases will be crucial to navigating the next internet era. Raising vital questions to identify high-quality versus low-quality information and ethical versus unethical online behaviors and communicating openly and respectfully is imperative to adapt to such a fast-paced digital world. The virtual universe will constantly test our perspective, integrity, purpose, and autonomy. Exercising critical thinking will be necessary to thrive in a dynamic environment of mixed physical and virtual interactions. While machines and algorithms will try to drown us with massive amounts of irrelevant information without filtering the flow of data, the onus is on us to rationalize, evaluate, and judge its pertinence to avoid being overwhelmed or manipulated.

The next era of the internet will create unlimited and unimagined possibilities for communications, e-commerce, personalized education, tailored financial services, and healthcare. It will redefine how people connect, consume, share content, and engage with companies. The changes we are experiencing today are still in the internet's infancy; the transformations of the digital landscape will be fascinating and scary at the same time due to the speed and breadth of disruptions. Solid policies and regulations must protect people's data and human autonomy for the internet to grow fairly and ethically. Companies need to approach the exponential growth of the internet with responsibility and ethics. If built ethically and safely, AI, machine learning, and social media platforms will connect people and the world in unprecedented ways by creating more human-centric, engaging, and immersive experiences rather than exacerbating polarization and divisiveness. We all need to play a more prominent role as global citizens to navigate virtual spaces responsibly. To build healthy relationships with technologies and the digital world, we must continually increase our digital literacy and align our digital footprint with our values and purpose. The next phase of the internet needs to augment our human capabilities and improve our well-being instead of harming our identities or influencing who we are as human beings. People need to be empowered by technology, not limited, manipulated, or threatened.

A 2050 Scenario: **Living Where You Love**

You live in Los Angeles, California, because you enjoy its diversity, cultural heritage, and big-city experience. Over the last decades, the city has become greener, more affordable, and more sustainable. All your favorite amenities are a fifteen-minute walk away. 3D printers built your home in less than four weeks, costing a fraction of the price of early twenty-first-century real estate and using significantly fewer resources. The house is eco-friendly; it generates, stores, and reuses its energy, water, and waste. Flexible structures expand the walls as the family grows. Digital technologies and virtual realities have made remote work effective; cities are now working hard to become more desirable as residents are chasing livability first, followed by jobs and salaries. Besides LA's energy and vibrancy, you appreciate the modernization and integration of mobility options such as buses, trains, and bike paths, as well as the common green and social areas where humans interact and talk with each other.

On November 23, 2050, you are preparing to fly to New York City to spend the Thanksgiving holiday with your family. Your flight from LAX spaceport to JFK spaceport is at 9:15 a.m. Your AI home assistant has booked your transportation to the spaceport based on the flight information stored in your digital wallet. Your transportation options range from electric trains and buses to flying drones. You chose a robotaxi, which has achieved full autonomy. Cars have become an entertainment center, an office, a classroom, or even a gaming room on wheels; you enjoy the ride while the vehicle does the rest. Your driverless car is waiting for you to leave your home at 8:30 a.m. Although you leave less than one hour before departure, there is no reason to worry about traffic; electric and full autonomous cars have not only eliminated transportation pollution and crashes, but they can take a high-speed hyperloop corridor with no intermediate stops, reaching 150 miles per hour. Arriving at the spaceport at 8:50 a.m., you reach your gate, ready to board. The entrance door into the spaceport lobby has already scanned your biometrics and verified your identity using a face scan, checked you in for your flight, and allowed you to avoid the stress of security belts and screening. By nine o'clock, you are sitting in your seat, waiting for takeoff. A trip that used to take nearly five hours is now over before the first episode of your favorite series ends—a spaceship can fly you from Los Angeles to New York City in less than thirty minutes.

◆ ◆ ◆ ◆ ◆ ◆ ◆ ◆ ◆ ◆ ◆ ◆ ◆ ◆ ◆

By 2050 IoT devices, face recognition, AI, cloud computing, autonomous vehicles, space tech, and high-speed networks will make this imagined scenario possible in many parts of the world, improving everyday life with accessible, safe, and enjoyable technologies. The future of cities will create more sustainable and fulfilling living methods for their residents. City planning will need to put people and the environment at the heart of all efforts. Human development and well-being need to be the primary measure of progress rather than economic growth. Over the next thirty years, many parts of the world will face rapid urbanization, testing cities' ability to support current and new residents safely, equally, and sustainably. There will not be a one-size-fits-all approach, as each city will develop differently due to its unique economic, social, and cultural aspects. Their ability to expand and adapt to new conditions will rest on four core areas: sustainability, mobility, accessibility, and transparency.

Urban spaces will be home to nearly 70 percent of the global population by 2050, increasing from 55 percent today. Ninety percent of this projected urban growth will occur in Asia and Africa. For centuries, cities have been the backbone of global economic growth, culture, and heritage and the hub of some of humanity's most significant innovations. Although cities will continue to generate worldwide wealth and prosperity in the next thirty years, rapid urbanization will exert additional pressure on an already challenged ecosystem, affecting the climate, energy production, healthcare support, living conditions, and transportation.

Though cities occupy only 3 percent of Earth's land, they consume over two-thirds of the world's energy and account for more than 70 percent of global carbon emissions. Therefore, cities need to become smarter, more self-sufficient, and more sustainable to support urban sprawl, for example, designing resilient solutions to keep people safe from climate change effects. Focusing on strategies to mitigate climate impact is necessary mainly for individuals who live in coastal areas. Over 570 cities will face a projected sea-level rise of at least half a meter by 2050, putting over 800 million people at risk from rising seas and coastal flooding.

Cities can play a vital role in the fight against global climate change. Sustainable strategies to reduce pollution and improve access to clean water and sanitation, new forms of housing, more efficient mobility, and more sustainable infrastructure must become top priorities for city governments and local businesses if they aren't already. As urban populations and economies grow in many parts of the world, the shift to renewable energy is critical to help mitigate environmental impact. The US Energy Information Administration estimates that renewables like solar and wind will be the "primary source for new electricity generation" by the middle of the twenty-first century. In the coming years and decades, buildings, factories, and residences will need to become self-sufficient by leveraging solar panels and energy storage systems to capture as much energy as they consume.

From today until 2050, new technologies and transportation models will reinvent the transportation sector. This transformation will significantly redefine how people and goods move from one place to another. With cars sitting idle over 90 percent of the time, owning one and paying for insurance will be redundant in the future. Instead, ride-hailing, integrated public transportation, and mobility-as-a-service apps—digital platforms that enable people to plan, book, and pay for multiple mobility services—will offer more connected and diversified options for citizens to move around.

The International Energy Agency (IEA) has mapped a scenario on how the global transportation sector can reach net-zero emissions by 2050. In "Net Zero by 2050: A Roadmap for the Global Energy System," IEA outlines steps including an aggressive uptick in sales of electric cars from around 5 percent of global car sales today to over 60 percent in less than a decade. By 2035 nearly all cars sold globally must be electric, and by 2050, nearly all heavy trucks sold must be fuel cell or electric. Achieving these milestones will entail more attractive incentives for people to upgrade their vehicles to electric models, electrifying ride-sharing cars—since they tend to drive more miles— and more convenient access to charging stations.

Up to 40 percent of inner-city traffic involves searching for a parking space. In the upcoming decades, developing a more efficient mobility ecosystem, including underground tunnels for autonomous vehicles to travel at high speeds, integrated public transportation, and flying cars and drones, will be vital to relieve traffic in many cities around the world. These advances will also allow parking spaces and garages into green spaces in city centers. With more intelligent and more sustainable vehicles on the road alongside more integrated urban mobility, many cities worldwide could experience a significant reduction in accidents, congestion, and emissions in the decades ahead.

As the urban population expands, the need for more adequate and affordable housing worldwide will increase. Today it's estimated that over one billion people are living in slums or informal settlements, with 80 percent concentrated in East and Southeast Asia and sub-Saharan Africa. By 2030, three billion people will likely have inadequate shelter. New forms of housing need to address challenges such as affordability, space limitations, and sustainability while boosting the quality of life for city dwellers.

More affordable housing, including high-rise apartment buildings, will be critical in the coming years. Countries around the world can learn and apply the successes of Singapore, a strong example of housing development. Singapore has created a public housing program known as the Housing & Development Board (HDB) that provides sanitary living conditions to replace unhygienic slums and crowded squatter settlements. Today HDB flats are home to over 80 percent of Singapore's population. Besides quality and affordability, residences will need to become eco-friendly to support a growing urban population; for example, they need to include rooftop solar power and water collection systems to make them self-sufficient. In more dense locations, new types of residences, such as shared housing, can mitigate space limitations, lower maintenance costs, and foster more interactions as residents share multifunctional spaces like kitchens, dining, and entertainment rooms. Soon more high-tech and futuristic housing architecture like 3D-printed homes will emerge in many parts of the world as a sustainable alternative to the current housing challenges, especially as they require a fraction of the time and cost of traditional construction. ICON, a pioneer in 3D-printed homes in the United States, in partnership with developer 3Strands, recently completed four homes in Austin, Texas. According to ICON, construction using 3D printing can be up to 30 percent cheaper and several months faster than conventional construction. Mighty Buildings, another 3D-printed home builder, claims the production process eliminates 99 percent of construction waste.

Cities will need to improve education systems to ensure long-term economic progress; by providing high-quality education to all citizens, cities will have a chance to reduce inequalities and enhance people's well-being. By adopting digital technologies, cities can provide more robust and inclusive education for students at school and at home. Accessible and quality healthcare solutions will also be vital for driving equality and inclusiveness within cities. Here, technology will also play an important role. For example, remote patient monitoring, robotics, and AI-powered analytics can reduce the healthcare burden in many cities worldwide. IoT devices such as smartwatches and heart-rate and glucose monitoring systems can measure people's vital signs in a real-time while directly connecting to healthcare centers and doctors. They can identify and transmit insights about health problems or variances, alerting the patient and the doctor if an intervention is needed. Cities facing doctor shortages can use telemedicine to cover the gap. If deployed safely and ethically, these technologies will help increase healthcare access and affordability in many countries around the world.

The expansion of community apps will help residents become more engaged with their local community, explore activities in their cities, and share resources. In addition, social media networks will increasingly become vital tools to enhance social engagement by enabling the population to dialogue with local officials directly. As city leaders and residents embrace digital technologies, citizens could collaborate on city planning and strategies instead of leaving these in the purview of city administrators and companies. As cities become more connected and automated, residents will have more significant opportunities to access real-time information to track and understand their city's progress toward critical priorities and projects. Transparency will be the foundation of the smart cities of the future. More facts available at the hands of people will transform public and city administration in the twenty-first century.

The human population will rise in the coming years, but the population of machines, robots, and sensors will outnumber them by the trillions within city centers by 2050. Technologies such as AI, machine learning, IoT, and face recognition will make cities more innovative, sustainable, and inclusive. The benefits of becoming a smart city are limitless. For example, those cities can use digital technologies to monitor air quality, manage energy and water consumption in buildings and common areas, estimate greenhouse gas emissions, monitor real-time waste management, and track roads to evaluate traffic conditions. Face recognition, if done with proper oversight and regulation, can improve safety by preventing crimes and boosting incident response times. For example, authorities can use this technology to identify a fraudulent criminal, fight child abuse, or help find missing persons. Virtual reality and augmented reality will transform cities into immersive engagement centers and create new possibilities for people to virtually engage with local arts, recreation, and tourist attractions.

Deploying digital technologies to streamline public services can help decimate cities' inefficiency and bureaucracy while providing more agility and improving user experience. City officials can learn from Estonia, known as a digital republic. A small country in northern Europe, it has digitized 99 percent of its public services. Estonians enjoy e-voting, digital IDs, e-prescriptions (99 percent of medical prescriptions are handled online), and remote verification for notaries. In contrast, the average American taxpayer with no business income spends eight hours filing a tax return. In Estonia, this process takes just five minutes. Estonia only requires physical presence for two civil actions: marriage and divorce. The country offers one of the most comprehensive governmental online services in the world. Using technologies such as blockchain, they ensure the integrity of data stored in government repositories and protect its data against insider threats.

The accelerated growth of city centers has contributed to the climate crises, inefficient mobility, growing inequalities, and housing shortages in many countries around the world. Cities need to be an integral part of solving environmental, societal, and economic challenges. They need to become self-sufficient by producing their resources, prioritizing circular solutions—including recycling, restoring, and reusing water, materials, and energy—and expanding electric transportation systems. These are pivotal steps to creating more sustainable communities and a more integrated urban life for the future. In addition to measuring thriving cities by their economic growth, we must also consider their residents' overall satisfaction, including happiness, inclusiveness, safety, and longevity.

A 2050 Scenario: **Feeding Ten Billion Human Beings**

You find yourself craving a pizza, not only because it's one of your favorite meals but because you can still remember the taste and smell of the pizza you had on your last vacation in Naples, Italy, known as the home of pizza. It's Saturday evening, and your always-on AI home assistant identifies your desires and preferences by simply monitoring your thoughts, proactively suggesting a homemade pizza. It tracks that the last time you had one was weeks ago. After your assistant recommends a few recipes, you choose an authentic margherita pizza, adding some prosciutto as an appetizer. However, you realize that you don't have the essential ingredients at home. Nothing to worry about! A modular vertical farm less than a mile from your home produces San Marzano tomatoes and fresh basil. The ingredients are organically grown with zero pesticide and little water. AI systems and machine learning control temperature and irrigation. The prosciutto has been safely cultivated in bioreactors in less than a month, as meat from real animals is rare nowadays. Your AI home assistant places your order directly with the grocery store. There, a humanoid robot receives it and gathers the products while 3D printers create the delicious and non-GMO prosciutto. The ingredients are delivered to your door by autonomous flying drones fifteen minutes later. Your household robot receives the package, washes and slices the products, then bakes the freshest margherita pizza. Your only job is to open a wine bottle and enjoy the moment.

Some of these technologies exist today, but they are not yet widely available or fully integrated into food systems, homes, and everyday life. However, we can expect to see them soon to fulfill the growing demand for food over the next thirty years.

The environmental impact of food production today is staggering. Globally, food and agriculture account for 26 percent of greenhouse gas emissions and 70 percent of freshwater use. Ninety-four percent of mammal biomass—excluding humans—is livestock, representing more than 50 percent of emissions generated by the food system either through land use, animal feed, land conversion, or methane production. Moreover, today 50 percent of the planet's habitable land is used for farming, compared to only 4 percent nearly a thousand years ago.

The Food and Agriculture Organization (FAO) of the United Nations estimates we will need 60 percent more food to feed a world population of nearly ten billion people by 2050. Besides increasing food production to satisfy a higher demand, we need to produce food using fewer natural resources and less environmental impact. It's also crucial to urgently address the disparities caused by malnutrition. According to indicators from the FAO, nearly 2.4 billion people lacked access to adequate food in 2020. In contrast, almost 40 percent of adults are considered overweight, and 13 percent are obese. Moreover, more than half of the global population lives in countries where complications from overweight and obesity kill more people than underweight.

As the global population continues to expand, integrated efforts and worldwide coordination from key stakeholders—governments, farmers, bioengineers, retailers, businesses, and citizens—will be required to ensure there will be enough healthy food to feed all human beings by 2050. The food industry is already making significant preparations to ensure that more efficient food systems are ready to fulfill 2050's demands for a higher quantity and quality of food. However, an important question remains: how will we feed ten billion people sustainably and avoid degrading the planet further? Tackling critical food challenges will require agricultural modernization, more efficient food waste management, and more sustainable and healthier food solutions, such as plant-based and cultivated meat products. As the end-users of the food ecosystem, we will play a critical role in influencing industry behavior and production toward sustainable, healthier, and more ethical practices in the years ahead. We can contribute by becoming more aware of environmental issues and adhering to healthier and more sustainable diets. We must rethink our daily habits and balance them. Our choices can improve our overall well-being and help shape the food industry's future.

Nearly 30 percent of the global population goes to bed hungry. Yet, each year, an estimated one-third of all food produced—1.3 billion tons, worth around $1 trillion—is spoiled due to poor transportation and harvesting practices, according to the United Nations. In developed countries such as the United States, most of that waste occurs in homes, restaurants, grocery stores, and food service companies. While due to unreliable storage or transportation, food is often lost between the farmer and the market in developing economies. Eliminating food waste and loss could help mitigate hunger in many countries. Many companies are developing solutions to address food waste inefficiencies; for example, California-based Apeel adds a layer of tasteless, odorless, plant-based protective coating on the surface of fruits and vegetables. This process helps keep moisture inside and oxygen outside, which means produce lasts twice as long. Hazel Technologies, a Chicago-based company, has developed a package insert that releases vapors that can triple shelf life by slowing aging in produce and preventing decay.

Digital technology is revolutionizing the agriculture industry by automating critical farming processes and empowering farmers to make data-driven decisions. According to a Credit Suisse report, developments in AI, drones, autonomous machinery, and intelligent irrigation systems could yield productivity increases of an estimated 70 percent by 2050. The convergence of these technologies enables more precise and sustainable farming practices by measuring, mapping, and managing real-time data on crops, livestock, and weather to increase crop yields while lowering production costs. Robots in farm activities are helping with fruit and vegetable harvesting. Autonomous tractors, powered by sensors, Global Positioning System (GPS), and AI, can harvest, plant, and spray. Drone footage can precisely identify plant disease or pest infestation locations, monitor livestock and irrigation systems, and create a more precise map of property boundaries. With more data, farmers can make more effective decisions; this enables food production to use fewer resources and reduce environmental impact. "Daily satellite imagery enables real-time decision-making at a global scale and at the local farm level. Farmers can grow more with less—optimizing inputs, reducing nutrient runoff, and efficiently scaling regenerative practices like no-till [plowing has been proven to release carbon from the soil] and cover cropping. Downstream consumer goods companies can improve supply chain transparency and validate zero-deforestation commitments," says Robbie Schingler, co-founder of the aerospace engineering firm Planet.

Two out of every three people will live in urban areas by 2050. More efficient and sustainable food production and supply-chain operations will be vital to support the rapid urban expansion. As agriculture already consumes half of the habitable land on the planet, one of the alternatives is to bring the farm closer to city centers and citizens' plates. Vertical farming, or simply growing crops indoors and vertically within urban spaces, is gaining traction as a critical solution to solve food production and supply chain challenges. Vertical farming uses aeroponics and hydroponics techniques to replace soils. And AI, robots, IoT, and big data create a controlled environment, efficiently managing light, humidity, and temperature.

Vertical farms can grow various crops, including tomatoes, lettuce, herbs, and flowers, using 95 percent less water than conventional farming while yielding up to 390 times more produce per square foot. According to a report from Credit Suisse, by 2050, vertical farming could be the source of 80 percent of the food consumed in urban areas. In the years ahead, buying fresher and healthier products will become more convenient thanks to companies like AeroFarms, a pioneer in large-scale indoor vertical farms. AeroFarms uses zero chemical pesticides, 99 percent less land, and 95 percent less water than traditional farmers. Infarm, a modular farming concept available in some grocery stores, allows consumers to pick the produce themselves. Their technology is "helping cities become self-sufficient in their food production, while significantly improving the safety, quality, and environmental footprint of our food."

The World Resources Institute estimates that global demand for meat could grow by 88 percent between 2010 and 2050, putting enormous pressure on forests and biodiversity. In contrast, if meat consumption in high-consuming countries declined 50 calories a day or 1.5 burgers per person per week, it would reduce the need for agricultural expansion and the associated with deforestation, even in a world with ten billion people.

Many startups are developing plant-based meat alternatives for healthier options and preserving our global environment. This trend will accelerate in the coming decades. For example, Impossible Foods' plant-based burgers use 87 percent less water, 89 percent less greenhouse gas emissions, and 96 percent less land than beef. Beyond Meat's plant-based burgers offer protein levels greater than or equal to their animal-based counterparts with no cholesterol, antibiotics, or hormones. By 2024 the global plant-based meat market is forecast to reach $24 billion, expecting this to grow to $85 billion by 2030. Most plant-based meat companies have replicated ground beef, chicken products, and flavored sausages.

Cultivated meat is another sustainable alternative to conventional meat growing in popularity; it offers the nutrients to meet consumer needs while reducing the impact on the planet.

Cultivation is the process of growing meat in a laboratory, which involves extracting stem cells from a live animal and growing them in nutrient-rich conditions. The cells grow in bioreactors at high densities and volumes, similar to bovine growth. While it takes a few years to grow a cow, scientists only need a few weeks to mature the cells that comprise whole cuts of steak; plus, there are no slaughtered animals. For cultivated meat to become a reality in the future, it will need to match or exceed the taste and texture of conventional meat at a competitive cost. Depending on consumer acceptance and price, the market for this meat alternative could reach $25 billion by 2030. One positive sign is that production cost has dropped significantly over the last decade: in 2013, the first cultivated meat cost around $300,000 a burger. By 2016 Upside Foods—previously known as Memphis Meats—had produced a "cultivated meatball" for around $20,000 per pound. In 2019 Future Meat reduced chicken production costs to $150 per pound and beef to $200 per pound. By 2021 the same company has produced four ounces of cultivated chicken breast for less than $5.

More than seventy startups are working to throw cultivated meat onto our grills in the coming years. Besides Future Meat and Upside Foods, an Israeli startup named Aleph Farms announced early in 2021 that it had produced the first lab-grown ribeye steak using a proprietary bioprinting process. This advancement will open the doors to making any other cut of meat. And the cultivated meat market is not limited to beef or chicken—it's expanding to new horizons, including seafood. BlueNalu has plans to introduce mahi-mahi as its first cultured seafood item, while Finless Foods has cultivated tuna made from actual tuna cells, featuring the same traits as wild-caught fish.

Instead of buying meat from the butcher, you can expect to be printing any type of cultivated steak cut or seafood from the comfort of your home in 2050. 3D printers will replace chef's knives; these printers will print and cut marbled steaks, mimicking the textures and flavors of conventional meat. As a result, there will no longer be any need to slaughter an animal for food. Additionally, a filet mignon or a bluefin tuna will likely cost no more than a chicken nugget.

The future of food is promising. However, society will only benefit if the food industry becomes more sustainable, efficient, and modern—from precision farming to transportation to waste management. More intelligent agriculture, vertical farming, plant-based and cultivated meats are examples of robust solutions to mitigate the shortfall between the amount of food we produce and the amount needed to sustain the world by 2050. In addition to these solutions, at the individual level, by increasing people's awareness about climate change and adhering to healthier diets and habits, we have a better chance of mitigating the urgent climate crisis.

The next waves of change will offer **new opportunities** for the world to prosper more **sustainably and equally.** However, they will also bring challenges and **new tensions that will test our ability to reinvent ourselves** and adapt to unpredictable scenarios.

2 Waves of **Change**

In the next three decades, disruptions from rapid technological innovations, shifting demographics, climate changes, and inequality gaps will shape the world's future. Already these four waves of change are transforming every aspect of our lives. As we look ahead, the accelerated pace and intersection of these trends will offer new opportunities for the world to prosper. However, they will also bring challenges and new tensions that will test our ability to reinvent ourselves and adapt to unpredictable scenarios.

The Technology **Wave**

The breadth and speed of technological innovations will increase in the years ahead, significantly transforming every aspect of human activity, including how we work, learn, play, and interact with people and machines. In the next thirty years, technologies will evolve rapidly, creating more disruptions and transformations than humanity has experienced in nearly three centuries. As a result, many solutions to global challenges, such as access to quality education, affordable and personalized healthcare, sustainable urbanization, and slow climate change, will rely on breakthrough innovations over the next few years.

Technological advances are moving faster than ever; what seems to be a breakthrough invention today could become obsolete tomorrow due to innovative ideas that are being developed faster and cheaper. The power and speed of computers, measured by calculations per second, have increased trillions of times over the last thirty years, resulting in significant efficiency gains and cost reduction. Since the beginning of the century, the price of cable broadband service has declined by 98 percent, allowing us to access higher-speed internet to work, communicate, play, and entertain ourselves online. Connectivity has exploded in the last fifteen years; in 2005, there were nearly 1 billion internet users worldwide, while today, there are 4.7 billion. Imagine the possibilities for collaborating, selling, buying, creating, networking, and learning when nearly 100 percent of the global population is connected to the internet by 2050.

In 1965 Gordon Moore, co-founder of Intel Corporation, defined the growth that would set the pace for our modern digital revolution as Moore's law. In his law, the number of transistors on a microchip doubles every two years, though the cost of computers is halved; this means that we can expect the speed and capability of our computers to increase every couple of years, and we will pay less for them. Another principle of Moore's law asserts that this growth is exponential. In biotechnology, the first human genome accurately sequenced was completed in 2003 and cost about $100 million. That cost dropped to less than $100,000 a decade later. Now it's less than $1,000. Research in genome sequencing technologies is not slowing; as the cost of sequencing a human genome continues to decrease, doctors will have an increased ability to obtain information of medical value for future care. This information can aid in identifying genetic variants with potential diseases, allowing doctors to act preemptively. In the next thirty years, genome sequencing and editing technologies could help prolong longevity indefinitely and enable us to live healthier lives. In the transportation industry, the cost of batteries for electric vehicles is declining markedly; a battery with the capacity of one kilowatt-hour (kWh) cost $7,500 in 1991. It dropped to $1,100 per kWh in 2010 and fell further to $137 per kWh in 2020. At this level, the battery cost for an electric car with a 75-kWh battery in 2020 was around $10,300. It would have cost more than half a million dollars for a single battery thirty years ago. Reducing battery costs is one of the critical factors to speed up the switch to electric vehicles.

Over the last century, it took decades for some technologies to reach millions of people. Today digital technologies and social media platforms are reaching global audiences within days as connectivity increases and more people and businesses are accessing the internet. It took the automobile industry sixty-two years to get fifty million users; mobile phones reached that same number in only twelve years. The internet reached its first fifty million users in seven years, while Facebook reached that number in three years. Twitter only took nine months. Since 2020 the adoption of digital technologies has accelerated as companies had to quickly find ways to continue operating and enable employees to work remotely during the COVID-19 pandemic. According to a McKinsey survey, companies have accelerated the digitization of their customer interactions and internal operations by three to four years.

The exponential growth of AI, machine learning, robotics, IoT, cloud computing, virtual reality, and many other technologies has reshaped the world in recent years. However, the convergence of seemingly unrelated and independent fields is rapidly disrupting products, industries, business models and reinventing our lives in unanticipated ways. As Peter Diamandis and Steven Kotler state in their book *The Future Is Faster Than You Think,* "these waves [of exponentially accelerating technology] are starting to overlap, stacking atop one another, producing tsunami-sized behemoths that threaten to wash away most everything in their path."

Our smartphones have absorbed the functionality of a telephone, camera, music player, and digital assistant powered by intelligent operating systems into one powerful device in a short period. The IoT—wearables, smart home, city, and industry devices—combined with AI, big data, and 5G networks will serve as the foundation for a hyperconnected world in the twenty-first century. While there were ten billion IoT devices worldwide in 2018, this number will reach sixty-four billion by 2025 and possibly many trillions by 2050. IoT's ability to gather and analyze real-time information and generate massive amounts of data will let us learn more about our bodies, cities, traffic, homes, and health.

Today's AI still has fundamental limitations and narrow capabilities. It can solve specific problems and tasks. However, as more data is collected, analyzed, and fed to those systems, they will become better at identifying patterns, optimizing human decision-making, and accurately predicting future outcomes. Some experts believe AI may match or exceed human capabilities and intelligence in the future. Artificial general intelligence (AGI) aims to create systems as intelligent and flexible as the human brain and can perform any task that a human can do. This fully realized vision of AGI does not yet exist. AI still has a long way to go before anything like human-level intelligence is achieved. "The absence of common sense prevents an intelligent system from understanding its world, communicating naturally with people, behaving reasonably in unforeseen situations, and learning from new experiences," says Dave Gunning of the Defense Advanced Research Projects Agency. "This absence is perhaps the most significant barrier between the narrowly focused AI applications we have today and the more general AI applications we would like to create in the future."

Although AI systems have delivered things that we might not have thought possible a decade ago, the science fiction we see in some Hollywood movies is still far away from becoming a reality. These systems are increasingly better at replicating cognitive abilities, but they still lack essential human traits such as consciousness, self-awareness, and judgment. By 2050 machines will be faster, more powerful, and more accurate, but our unique human abilities such as intuition, emotional intelligence, collaboration, imagination, and the ability to anticipate changing situations will allow us to navigate and overcome future disruptions and uncertainties. Unlike AI, machines, and robots, we can understand how the world around us functions. We can learn through lived experience and continuously adapt our behaviors to novel contexts while evaluating ethical implications.

The convergence of AI, machines, and robots will transform the nature of work in the next thirty years. By 2050 the breadth and speed of the successive waves of automation will radically alter the way we work, making today's workforce almost unrecognizable. As a result of automation, many jobs will disappear, yet we will see new job opportunities created to replace them. Historically, technology has been a net job creator. While there were significant concerns about job decimation with the rise of the internet, a study conducted by McKinsey Global Research found that the internet created nearly three new jobs for each one lost. Previously unimagined jobs emerged in entertainment, social media, retail, science, technology, and transportation, to name just a few. As we navigate the next waves of technological disruption, 50 percent of today's work activities could be automated by 2025 due to next-level process automation and virtualization. Automation will continue to drive economic growth via increased levels of productivity and efficiency. However, as machines take over tasks previously executed by humans, workers from every industry must constantly acquire new skills to avoid displacement. Humans and machines will be increasingly working in tandem in the fast-approaching future.

This collaboration will define the future of work.

We are increasingly digitizing many aspects of our lives through the widespread use of the internet. Instead of buying CDs or DVDs, we now subscribe to cloud-based apps to listen to millions of songs, and we stream thousands of movies anywhere and anytime. Most of our physical meetings, education, and training are shifting to digital formats using videoconferencing and collaborative platforms. We can exercise using bikes and interact with personal trainers on the other side of the world. We can meditate utilizing the help of wellness apps. We can schedule an appointment and talk with our doctor from anywhere by using telemedicine, healthcare apps, and wearable devices to monitor our vital signs. Physical money is disappearing as digital banks and cryptocurrency powered by blockchain technologies become mainstream.

Today we need to connect our smartphones or computers and look at a screen to access the digital world. Tomorrow, the delineation between online and offline will be a lot harder to describe as AI, virtual reality, IoT devices, and sensors come closer to our body and mind. These technologies will create more immersive, highly personalized, and collaborative digital experiences.

The first commercial computers developed in the last century were the size of rooms and weighed tons. Unless you had a cargo plane or a truck, you would not have been able to move them around. Personal computers and laptops allowed for greater mobility. Today we carry our smartphones in our pockets, allowing us to connect continuously with the online world. Soon technology will move from our hands to our faces as virtual reality devices will enable us to experience real presence in virtual spaces. Immersive experiences will reinvent education, healthcare, manufacturing, entertainment, finance, and retail. Digital spaces will create novel ways to interact, play, and connect with friends, family, and coworkers, significantly redefining social and business interactions.

By increasingly digitizing our experiences, we will have new ways to seamlessly access real-time information, entertain ourselves, and connect with billions of people worldwide in more engaging and present forms. Virtual spaces certainly create exciting new ways for businesses to communicate with customers, develop immersive training platforms, and streamline processes like manufacturing. Virtual environments offer the potential to significantly redesign the future of work, making it more productive and interactive. For example, Microsoft's Mesh is a mixed-reality platform that promises to make remote interactions feel more lifelike. Alex Kipman, a Microsoft technical fellow, describes the technology: "You can actually feel like you are in the same place with someone sharing content or you can teleport from different mixed reality devices and be present with people even when you are not physically together." For example, people can increase the sense of presence and have more engaging meetings than simply videoconferencing. Doctors, firefighters, or police practicing dangerous or life-threatening situations in virtual settings can be more effective than relying on theory. Engineers, designers, and product managers will immerse themselves in virtual scenarios, projects, and training to amplify real experiences. By immersing themselves in digital spaces, they can design, iterate, and test concepts and receive feedback in real-time.

The concept of the metaverse is receiving heightened attention nowadays. Matthew Ball, a venture capitalist, has described it as "a sort of successor state to the mobile internet" — a framework for a highly connected life. The idea is to blend the digital and the physical world, creating digital spaces to bring people together to play, socialize, and work. The concept behind the metaverse is an extended reality combining augmented, virtual, and mixed reality in new ways for social interaction and business engagement. The full potential of the metaverse does not yet exist. However, the technology is expected to be as transformative as the mobile phone. As we spend a lot of time looking at screens and navigating the digital world every day, the delineation between physical and virtual lives has blurred. From working, learning, gaming, banking, chatting, to buying and selling digital goods using non-fungible tokens (NFT), in some form, the metaverse is already here.

Like any emerging and disruptive technology, it brings challenges to safety, privacy, and security. Safely engaging in the metaverse will require constructing a trusted ecosystem where people are protected from online harassment, assaults, cyberbullying, and hate speech. As our energy and attention are increasingly moving away from the physical to the virtual world, we need to leverage AI, social media platforms, and virtual realities as a vehicle to take us in the direction we want for our future. These technologies need to augment our intelligence and help us experience the world in more beneficial, immersive, and engaging ways, rather than creating a parallel virtual life where our real-life becomes less important. Increasing our presence in the digital world cannot remove the responsibility and accountability we have over our lives, obligations, and well-being in the physical world.

The Climate **Change Wave**

Climate change calls for urgent action. We are experiencing rising sea levels, extreme weather events, melting glaciers, and hotter temperatures in many parts of the world. Although climate change can affect all countries, some effects will be felt disproportionately by more vulnerable and less prepared regions. The transition to renewable energy, smart agriculture, deforestation, and the switch to electric vehicles are some solutions that will help mitigate the environmental impact in the twenty-first century.

Evidence from recent decades demonstrates that we are in a

planetary emergency. We need to take urgent and significant actions now; otherwise, the effects of climate change will bring irreversible outcomes. According to the Global Trends 2040 report published by the National Intelligence Council, the past decade was the hottest on record; every decade since the 1960s has been warmer than the previous one. If nothing changes, global warming may surpass 1.5 degrees Celsius within the next twenty years while heading toward 2 degrees Celsius by the middle of the twenty-first century. As temperatures continue to rise, we will face more frequent and intense natural disasters, including hurricanes, droughts, and coastal flooding. In addition, some estimates suggest the sea level could rise to fourteen inches over the next two decades, creating challenges for coastal cities and islands across the world.

Biodiversity is declining faster than at any point in human history, risking food and health security while also undermining global resilience. Climate change could become one of the critical factors for emigration as people seek better and more sustainable environmental conditions. If we remain on our current path, only the wealthiest individuals will continue to hide in air-conditioned paradises far away from impacted zones. At the same time, the poor will be left exposed as they try to survive in harsh circumstances.

Avoiding a climate catastrophe will depend on collaboration among key stakeholders, including the private sector, governments, and citizens. How quickly they can act now will dictate how efficiently we can avert many life-threatening scenarios for humanity in the twenty-first century. Consumers and investors will increasingly pressure companies to keep their promises and accelerate their strategies and commitments to achieve net-zero emissions. In the coming years, achieving higher environmental, social, and governance (ESG) results will be as important as revenues and profits, if not more so, in the eyes of shareholders, employees, and consumers. Likewise, governments around the world will be held accountable for upholding the 2015 Paris Agreement, which attempts to halt the rise in global temperatures below 2 degrees Celsius and pursue efforts to keep it under 1.5 degrees Celsius.

Humans can be the help that Planet Earth needs with consistent coordination, transparency, and partnerships between the private and public sectors. These actions are essential to solving the climate crisis. However, to close the net-zero equation, we need to increase citizens' engagement. Along with technologies, companies, and countries, individuals are catalysts for positive change. So how can we become more involved to help fight climate change?

We can drive the changes we want to see by increasing our awareness about climate change and reducing our own impact. Our household activities are directly responsible for up to 20 percent of all greenhouse gas emissions and indirectly, through consumption, for up to 70 percent. We can start by adjusting our behaviors—revisiting our everyday activities and habits, including what we buy and where, and rebalancing our diet toward more sustainable and healthier food. By doing so, we can effect change at the individual level.

As we navigate a world where collaboration and inclusiveness will be more critical than ever, everyone needs to be part of the solution to achieve sustainability—meeting our own needs without compromising the ability of future generations to meet theirs. We need to think globally, act locally, and engage citizens for that to happen. The conversation needs to move beyond collaboration and agreements between companies, CEOs, governments, and banks. We need to make the climate change discussion more tangible and evident to everybody. One way to do this is to create an individual climate change score to track what people consume, drive, wear, and waste. People need to measure their impact on the environment consistently and accurately, as easy as accessing an online banking account. This score would help increase people's awareness of how their actions and everyday habits impact the environment by providing individualized data. Depending on their consumption, the score could be used as an indicator to aid them to access better loan rates, lower monthly utility bills, or even higher tax deductions. Increasing awareness about climate change and incentivizing people to change their habits and consumption behaviors is vital for a more sustainable planet.

The Demographics **Wave**

According to some projections, despite the global population growing slower, it will reach almost ten billion people by 2050. The global population increase will come primarily from developing countries. Ninety-three percent of the growth expected in the next three decades will occur in Africa (59 percent) and Asia (34 percent). By the middle of the twenty-first century, the population of India and China combined is likely to surpass 3 billion people. Africa's population will be nearly 2.5 billion people by 2050. In the United States, the population is expected to reach 398 million in thirty years, compared with 335 million in 2020.

The speed at which technology is allowing people to come together in the digital universe will remove boundaries between countries, generations, ethnicities, genders, and cultures. As more people get online, the future of the internet will be more diverse and representative of the global population. We will likely see more trends, businesses, and influences coming from developing nations. From now until 2050, we will build relationships and work in a more geographically dispersed, ethnically diverse, and digitally connected workforce than any previous generation. Digital technologies are increasingly enabling companies to find and hire talented resources and allowing their employees to work virtually from every corner of the planet. To flourish in a more integrated, globalized, and diversified world, our human traits will be essential to build trust, break down cultural barriers, and adapt to new cultures and local scenarios. The key is to remove ambiguity by effectively communicating and becoming an active listener, respecting different perspectives, and appreciating that different generations learn, engage, and work at different speeds and in different ways. We need to empathize with others and better understand local cultures and behaviors. Building these skills and increasing our social awareness is essential to grow as global citizens and embrace a broader range of cultural aspects.

The combination of smaller families and longer lives will result in an older global population in the twenty-first century. As longevity increases, the future will allow individuals to enjoy their loved ones and passions. By 2050 reaching and perhaps exceeding one hundred years will be a reality for many more people. However, longevity will bring challenges to many individuals and create financial burdens for governments as healthcare, pensions, retraining programs, and unemployment benefits will need to support the increasing demands of an aging population.

People over sixty-five years old are likely to approach 25 percent of the total population by 2040 in many older countries, including several advanced economies; in 2010, this number was 15 percent. As the population ages, individuals will likely need to work longer and reinvent themselves continually to support their financial needs over an extended lifetime. Today a sixty-year-old worker may be unwilling to explore different alternatives or transition to a new career as they march toward their long-awaited retirement. By 2050, in a world where AI, automation, and digitization will be commonplace and dictate the pace of transformations in the workplace, adaptation, flexibility, and lifelong learning will be necessary skills for enduring employment. Refusing to change and adapt to new scenarios will make workers redundant. Lifelong learning has no age limit, and in the future of work, its importance will be essential to keep up with technological changes and disruptions.

The Inequality **Wave**

The first three waves of change that are currently disrupting and transforming the world—climate change, technological revolutions, and shifting demographics—can exacerbate inequalities further by 2050. The income gap between the top and bottom earners continues to grow. The aggregate wealth of the world's richest 1 percent has risen nearly four-fold from $41.5 trillion in 2000 to $191.6 trillion in 2020, an increase from 35 percent to 46 percent of global wealth. Inequality gaps significantly impact life expectancies and future opportunities. For example, the difference in longevity at birth between countries with low and very high human development is nineteen years. People in more developed nations spend seven more years in school, reflecting critical gaps in access to health, education, and overall human development.

Digital technologies can help create a more inclusive, educated, and engaged global workforce, but it's critical to reduce the digital disparities in many parts of the world. For example, close to 87 percent of the population in developed countries have internet access, compared to only 19 percent in the least developed countries. Reducing this gap would create more social mobility and economic stability as more people in remote and vulnerable areas could access affordable and online education and pursue higher-paid jobs.

AI, machine learning, robotics, and automation will increasingly take over routine and manual work, making workers redundant for many existing tasks and jobs. Ensuring internet access and lifelong learning are available for a larger number of people is essential to equip them for a job or career change. The rapid expansion of automation may lead to lengthier unemployment, further increasing the unequal income distribution as workers need to develop new abilities to compete in other fields and industries. A more educated person transitioning to a new job or career may find other opportunities easier and faster. However, the rapid advances of AI, robots, and automation will challenge lower-skilled professionals to access jobs requiring higher-level skills and expertise. AI systems pose another threat as they grow considerably more complex and sophisticated. If trained with data sets that include biased human decisions or reflect historical and social inequities, they can negatively influence decision-making and outcomes. False and misleading judgments in sensitive areas, including credit applications, criminal justice records, and job applications, can significantly impact people's lives and reinforce gender, racial, or sexual discrimination.

The World Economic Forum has predicted it will take 268 years to achieve gender equality in economic participation, according to the "Global Gender Gap Report 2021." Radical and mindful actions need to address the gap in gender pay; otherwise, this will continue to be an issue for centuries to come. So how can we achieve gender equality in significantly less time? First, increase women's active participation in the workforce and ensure they have access to continuing education to develop abilities in growth areas such as science, technology, engineering, and mathematics (STEM). The growth in STEM jobs is expected to outpace non-STEM jobs in the coming years; increasing women's skills in those fields can help them compete for quality and well-paid jobs. In addition, preparing women to fulfill leadership roles is essential to boost their engagement, motivation, higher economic participation, and inspire future generations.

Inequality is a complex challenge, and the solution will not come solely from a technological, economic, or political approach; instead, it will require global coordination from key stakeholders, including governments, companies, international organizations, and individuals. We can be the glue that holds people and communities together. By promoting social cohesion, we can strengthen relationships and solidarity with others. As digital technologies and social media platforms remove the lines between the physical and digital worlds, we are increasingly becoming part of a much larger community. To mitigate disparities, we need to ensure this new global community progress without boundaries of ethnicity, gender, race, or sexual orientation. Tackling inequality issues, building trust, fighting discrimination, welcoming inclusive dialogues, eliminating biases, and embracing inclusiveness are vital ingredients to building social cohesion over the coming years. Our actions can contribute to a stable, harmonious community. Our job is to promote unity and focus on changes that can lead to global transformations while erasing political, social, and cultural polarization in both the physical and digital worlds.

Inequality should not affect anyone's destiny by 2050. Everyone must have equal access to the internet and digital technologies to access lifelong learning and expand their skills in the coming years. Those without a future-ready mindset will be left behind. Continued education can level the playing field, whereas a lack of personal initiative, human-centric skills, and financial literacy will only reinforce existing inequalities.

A 2050 Scenario: **A Fast-Changing World**

Noah is lucky enough to be surrounded by family members of numerous generations, from his 30-year-old daughter, Eva, to his 120-year-old grandmother, Maria.

Maria - Born in 1930, now 120 years old.

Maria was born in 1930 and grew up facing some of history's most disruptive events. Her childhood was clouded by the Great Depression as her family scrounged for every scrap of food; they lived dark days of hopelessness. Maria spent her preteen years fearing World Wars as they had ravaged what little she knew of the world.

Maria witnessed the transition from the industrial era to the digital revolution in the latter half of the twentieth century. Although social and political upheaval added turbulence, life progressed more predictably in the twentieth century. Basic education and hard work would undoubtedly yield a good living. People could raise a family and then enjoy retirement later in life. Her adult life moved along similar to that of her parents, without much change. Things evolved event by event. Everyone knew their neighbors, and communication was predominantly face-to-face. Global connectivity was not broadly available.

Noah - Born in 1990, now 60 years old.

Noah has experienced unprecedented breakthroughs in the political and financial arenas. But the most significant changes were driven by the rapid technological advances that have reinvented every domain of his life, personally and professionally. Unlike his grandmother, the waves of change he encountered occurred rapidly and globally. Since 2021, he has experienced the exponential growth of the internet, the explosion of social media networks and smartphones, and the rapid advances in AI, virtual reality, and IoT. Information and misinformation flew from anywhere, anytime, at unprecedented speed.

He is a graduate of an archaic educational framework created for the industrial era. His four-year college degree didn't secure lifelong jobs like his parents held, nor did he want to spend decades in the same career. Social cohesion, purpose, and meaning were the elements he valued the most throughout the last decades. Since he entered the workforce, no jobs have been unaffected by AI and automation, which pressured him to acquire new abilities and unlearn old skills to avoid becoming obsolete. Degrees or certifications no longer secured a job. However, his ability to adapt to unique circumstances, cultivate a positive attitude, and improve his human skills allowed him to experience many fulfilling jobs in different industries as business models continually evolved.

Noah grew up more environmentally and financially aware than his parents. He has given up car ownership, and today his commute involves a combination of flying cars and autonomous vehicles. He has navigated challenging times due to geopolitical tensions that increased global uncertainty. Unfortunately, physical wars haven't ceased in many parts of the world. Moreover, conflicts between countries and governments have also escalated to new battlefields; new cyber soldiers and armies have dominated the cyber arena as everything has gone digital.

Eva - Born in 2020, now 30 years old.

Change has accelerated for Eva. She skipped the smartphone era altogether as more immersive and engaging experiences in virtual spaces exploded over the last thirty years. The physical world has merged with the digital universe. Some people still cannot make sense of it or recognize what is real versus virtual, while others have decided to surrender the physical world and live in virtual castles, drive virtual cars, and create a virtual identity. Their bodies are still on Planet Earth, but their minds and personalities reside in a distant digital universe. Eva is fortunate to have embraced virtual spaces to collaborate and experience the world in exciting new ways. She shops, makes friends, and explores other continents in virtual spaces, but she values her life in the physical world more than anything. As social media platforms tried to permanently wire humans to a virtual world and dominate their attention completely, people's autonomy, data privacy, and security became even more vulnerable. But tracking people's information and collecting their data were the least of their problems; cybercrime erupted as people have been hacked or abused without knowing they were interacting with criminals, while some brands excelled at persuasion and influencing behaviors.

In addition to immersing herself in cyberspace, Eva is an early adopter of direct brain-computer connectivity; instead of using virtual reality glasses and lenses to connect with the digital world, she connects her brain directly to the internet. Things happened locally in the twentieth century and rapidly expanded globally in the twenty-first. Inspired by the rapid advances in space exploration, Eva is already thinking about becoming a multi-planetary citizen in the twenty-second century.

The **singular humans** imagine the future they want to build and pursue it passionately. **They embrace change, take risks, and don't feel paralyzed in the face of difficulties and unknowns.** On the contrary, their biggest fears are stagnation and normalcy.

3 The Singular Human of 2050

The acceleration of technological advances will directly impact our lives, education, and jobs. As our physical and digital lives merge, AI, machine learning, and social media platforms will increasingly test who we are as human beings, challenging our identities, values, beliefs, and what we stand for. By 2050 the world will be filled with people leading change and those unwilling to adapt to new scenarios. There will be visionaries and innovators, but there will also be laggards and people who will always be playing catch-up in learning and self-development. The people who imagine and chase their dreams will live a fulfilling life; in contrast, those floating through life without motivation or clear direction will be drawn into a life of defaults and standards. Some people will use technology and social media platforms to unite other individuals, cultures, and communities, but others will try to divide the world further. Some will use technology to improve their lives and augment their inherently human abilities. In contrast, others will try to compete against machines or become like them, consequently losing their ability to think for themselves and make sense of the world around them. Some individuals will create essential technologies with justice and fairness built into their algorithms, while others will only boost profits.

In a not-too-distant reality, people and technology will share space in society, especially in the workplace. Some people will anticipate and thrive in uncharted territories; others will be left behind. The secret to maintaining your distinctiveness is to equip yourself with human skills and remain flexible to adapt to a fast-paced environment.

The next waves of change will create ambiguity and uncertainty in the next years and decades. Nurturing human skills will be critical to responding quickly to the next tsunami of technological, social, and economic developments heading for us in the future.

Not too long ago, robots performed unsafe, repetitive, and monotonous tasks with less human interference, mainly in manufacturing. Today they are all around us: cleaning, flying, talking to us, helping us buy products online, delivering our goods. Robots are helping doctors and nurses care for patients, and in the future, they may even perform precise and complex surgeries. There are even humanoid robots resembling the human body in shape. In 2021 Elon Musk announced the Tesla Bot; he suggested, "It could run to the grocery store for you, but presumably it would handle any number of tasks involving manual labor."

Drones and autonomous vehicles are becoming essential technologies in agriculture as they can handle tedious and repetitive activities for farmers and provide massive amounts of crop and weather data. In addition, robots are quickly becoming essential resources in dangerous conditions for humans, such as wildfires, hurricanes, and floods. Smart thermostats, intelligent appliances, smart light bulbs, robotic vacuums, and digital assistants are transforming our homes into intelligent and always connected living spaces. Soon, robots will become integral household members with duties and responsibilities like cooking, washing, and gardening; some will have a mission to prevent loneliness and boost mental health as they serve as caregivers and companions for the growing aged population.

AI is increasingly allowing machines to take over tasks performed by human beings, including writing articles and stories, creating songs in various styles, interpreting laws, and forecasting financial performance—the list goes on. This trend is transforming how journalists, artists, marketers, lawyers, financial planners, and many other professionals work. By 2050 AI, chatbots, and humanoid robots will be commonplace. As these systems become more intelligent, human-machine interaction will feel more natural and seamless. As AI and robots increase their ability to understand human beings' emotions and preferences and anticipate their needs and behaviors, some people may prefer to engage with machines instead of other humans. Imagine robots and chatbots that engage with us emotionally and understand our moods, feelings, and motivators—an annoying call to the helpdesk will be as enjoyable as chatting with your best friend. The rapid developments in AI and robotics have put us on the cusp of a new automation age. Besides being faster and more accurate than humans, machines are less expensive to maintain; we need time off, get sick, and change jobs.

Until 2050, all jobs and tasks that AI, machine learning, and robots can automate will be eliminated or transformed into something completely new. The world in 2050 will be hyperconnected, and machines will be everywhere. In a world where machines will oversee our lives, take over jobs from humans, and try to hack our emotions, nurturing our human-centric skills will be necessary for survival. You can't focus on standardization and uniformity. Your human advantage comes from continued learning, flexibility, and creativity. To maintain your distinctiveness, you will need to focus on the abilities that machines find difficult or impossible to emulate.

Maintaining Your **Distinctiveness**

A singular human strives for uniqueness and originality. This individual focuses on long-term benefits instead of short-term distractions, confusion, chaos, and uncertainty. They imagine the future they want to build and pursue it passionately. They embrace change, take risks, and don't feel paralyzed in the face of difficulties and unknowns. On the contrary, their biggest fears are stagnation and normalcy. Uncertainties and unknowns motivate them to explore and build new possibilities instead of accepting old models or solutions. They repeatedly reinvent themselves to adapt to an ever-changing world. They anticipate changes and prepare for them before they become urgent or unsolvable. They do not panic in the face of challenges and new scenarios; instead, they assimilate and synthesize relevant information to make data-driven and trustworthy decisions. They understand that technology grows exponentially, while human brains like straight lines. They know that it's not about competing against technology but leveraging it to augment their capabilities and reach new people and knowledge. They don't focus solely on their actions or contributions—they are enablers. They share knowledge to empower others to do better and make learning accessible to all. A singular human is ethical, and integrity is their top priority.

By 2050 the demand for singular humans and their skills will be unprecedented. Companies will fight to attract and retain their uniquely human qualifications and attributes. A singular human doesn't want to become a robot. They favor creativity, adaptability, critical thinking, and problem-solving over outsourcing decision-making to machines. Instead of refusing to accept the inevitable advances of AI, machine learning, and automation, they leverage them to free up their routine so they can allocate more time and resources to developing new skills and preparing themselves for the jobs of the future. They are not afraid to make their own jobs obsolete—more fulfilling and valuable opportunities will arise due to their discoveries. They have a growth mindset and are lifelong learners by nature, driven by action and a curiosity to explore the unknown.

Developing and maintaining our uniquely human capabilities will enable us to adapt to unexpected situations and remain relevant despite the increased presence of machines around us. More importantly, doing so will protect our most important assets as humans: our identities, values, and personalities. If we maintain our human distinctiveness, we increase our chances to flourish in a constantly changing world.

Intentionality

Willingness, consistency, and ongoing learning will aid us in creating new possibilities and ideas for a more purposeful future. In contrast, inertia and lack of anticipation and preparation will inhibit people from flourishing in a fast-paced digital economy. Precisely predicting the future is not a viable option, so becoming intentional learners by continually seeking new knowledge is our best tool for building an extraordinary life. Intentional learners are constantly preparing for a world that will not stop changing.

According to the Organisation for Economic Co-operation and Development (OECD), more than one billion jobs—almost one-third of all occupations worldwide—are likely to be transformed by technology before 2030. Intentional learners understand most of the skills and abilities we have today will be irrelevant in the future.

Intentional learners imagine their ideal future and take deliberate steps to build it. The idea behind intentional learning is the practice of treating every experience as an opportunity to learn something new. Intentional learners take this approach in each situation, interaction, relationship, and embrace lifelong learning.

Their curiosity guides their lives, and learning is not a separate workstream or an extra effort—it happens almost unconsciously. In a world where technology will transform all jobs and industries, intentional learning will be essential.

Intentional learners are moved by their purpose as it guides their decisions, influences their behavior, and keeps them headed in the right direction. An intentional learner, driven by exploration and curiosity, does not become addicted to or persuaded by technology. As the internet allows people to access many new possibilities, intentional learners can further amplify their knowledge. They glean more from the digital universe and use it as an opportunity to build connections and continue developing their skills.

Dr. Carol Dweck, a professor of psychology at Stanford University, reveals that the ability to learn is not fixed; it can change based on our efforts. In a hyperconnected and digital world, where the waves of change will transform all aspects of our lives, having a finite or fixed mindset is a mistake—the answers from yesterday will not apply to the questions and challenges of an unpredictable tomorrow. Instead, we need to develop a growth mindset to unleash new opportunities and expand our human abilities. According to Dweck, the individuals with a growth mindset "tend to achieve more than those with a more fixed mindset (those who believe their talents are innate gifts). This is because they worry less about looking smart, and they put more energy into learning."

Intentional learners are not afraid to take risks because even if they fail, they learn what to avoid next time. They see fear as a significant barrier to personal and professional growth; they confront their anxieties by remaining curious and exploring new ideas. As the world is constantly under transformation, their best strategy is to actively seek out and listen carefully to feedback, positive and negative. Intentional learners welcome new approaches, and their ability to make connections between disparate concepts makes them singular.

Relationships

Singular humans value, respect, and are loyal to their friends and family. They also maintain healthy relationships with machines. Nurturing relationships with people allows them to grow personally and help others around them. They consider relationships a necessary foundation during challenging times. Positive and productive relationships give them a higher sense of purpose.

Although AI and social media platforms are intertwined in their daily lives, proposing what they should watch, listen to, buy, read, and who they should connect with, they limit the influence these tools have over their lives. They are aware that by allowing companies and their algorithms to monitor and access their personal data in exchange for access to services, they may lose their autonomy and sovereignty over their digital presence. Therefore, the relationship they have with machines is positive. Still, they don't want to be the content of social media algorithms as they constantly try to manipulate and influence everyone navigating their platforms. Instead, they access meaningful content and connect with people to augment their intelligence and knowledge.

As machines and algorithms will increasingly monitor our bodies, minds, and thoughts, we cannot allow them to know more about ourselves than we do. Allowing machines and devices to alert us about our health conditions, recommend movies, songs, and products, optimize our online searches, and personalize our education, can help improve our daily lives and save us time. But we should not build a relationship in which machines influence us in opposition to what we stand for, who we are, and how we want our lives to evolve.

Singular humans are aware of the harm and benefits of digital technologies and avoid building toxic relationships with them. Protecting who they are and their human capabilities are critical for them to continue thriving in a world where machines gain more space in people's lives and jobs. They understand that respect, transparency, trust, and open communication—characteristics not found in machines, as they don't have a sense of being—are the foundation of a solid relationship.

Imagine a situation where a kid struggles to build relationships in the real world with peers and decides to create a digital twin to simulate herself in the virtual world through an adorable avatar so she can experience the pleasure of having friends. In which environment will she prefer to live her life, a simulation of reality or reality itself? What about people's appearance? Will they choose their digital version's face, body, and clothes rather than their physical reality? Filters are already the new makeup, and feeds and stories are becoming the central place for people to show off who they are—or wish they were. As machines are increasingly present in our lives, those who can accurately represent themselves and build healthy relationships with technologies by using them to share their unique abilities, knowledge, and real-life stories, will transform the world into a better place.

The winners in 2050 will be the individuals who favor authentic human relationships over toxic and dependent ones with machines and screens. We must embrace digital technologies to make life easier, but we cannot depend on them alone. Building relationships is not about the number of "friends" on social media feeds; humans need meaningful and trustworthy connections to thrive. The success measures depend on the quality and duration of human relationships. Machines will be introduced into everything we do at all times. However, our most trusted human connections will always serve as the foundation of our lives, especially during times of uncertainty.

Mental Flexibility

Singular humans remain flexible to adapt to continuously changing and unplanned environments rather than maintaining regimented thinking or letting an uncertain future paralyze them. They don't allow frustration and disappointment to emerge in the face of the unexpected; instead, singular humans continue to pursue their personal goals and dreams even if they confront obstacles along the way. Mental flexibility will be necessary to ride out the transformations that will shake up our lives.

The key to navigating a volatile environment defined by constant technological revolutions, social disruptions, changing demographics, and climate change is adapting to different perspectives, appreciating diverse opinions, and stimulating our creativity. Protecting our mental health, particularly by managing higher stress levels, will be a priority in the twenty-first century.

In a world that will never stop changing, we must become better at sensing and evaluating what is negatively affecting our lives. We need to create new possibilities for a more purposeful future instead of adhering to the same standards and assumptions of the past, with little flexibility or allowance for change. The future will be unpredictable, and not knowing the answers should not hold us back. Instead, imagination, willingness to learn, and flexibility will help us navigate and overcome ongoing challenges and unforeseen scenarios. Cultivating a growth mindset and revisiting our perceptions, emotions, and intentions will allow us to be more resilient, creative, and confident to adapt and flourish in the future defined by many unanswered questions.

Humans have the unique power of using imagination and intuition for reasoning, translating knowledge into different contexts, and making decisions spontaneously. These abilities give us an advantage over machines as we can interpret new situations and rapidly pivot to a different scenario when we see a better fit.

Furthermore, the ability to interpret the world holistically will allow us to anticipate and prepare ourselves for the next waves of change and growing threats from AI and machines.

Persistence

Over the next several years, we will need to evaluate our long-term objectives frequently so we can take advantage of the new circumstances created by the latest technological, social, and economic advancements. Resetting some of our beliefs and assumptions will help us adapt to new scenarios and reduce the risk of being left behind by the trends transforming all aspects of our lives. The road ahead will be challenging and foggy but motivating due to the immense new possibilities created by digital technologies and the billions of new internet users around the planet. Chasing a purposeful life will demand determination and persistence in the face of adversity. However, as we navigate an era of frequent disruptions and limitless opportunities, trying harder and persisting without a clear objective will not necessarily be the most optimal solution to achieving your goals. The key is accepting that we are always in progress.

Singular humans persist yet remain flexible and open to new directions. They are fully aware that persistence allows them to capture the most fulfilling and meaningful opportunities in the future. They consistently focus on their long-term objectives and purpose, particularly in times of difficulty and unexpected disruptions. Persistence is their best shot in an ever-changing and complex world.

By embracing failures as lessons learned, they recognize their limitations and frustrations and approach each wrong turn as an opportunity for self-development and personal growth. In the twenty-first century, the recipe for success will be developing and recycling human abilities to turn uncertainties and obstacles into novel possibilities. Dr. Susan David, a renowned psychologist, discusses the importance of becoming more emotionally agile in her book *Emotional Agility.* She highlights the importance of navigating life's changes with self-acceptance and an open mind. David states, "people who see themselves as agents in their own lives are more open to new experiences, more willing to take risks, more persistent, and more resilient in rebounding from failure. [. . .] Emotional agility can help one develop grit since it allows us to unhook from difficult emotions and thoughts, manage setbacks, and identify our values so we move toward a long-term goal worth pursuing. But it also allows us to let go of those goals once they no longer serve us."

Singular humans are unique because they trust themselves and have a sense of control over their lives. These characteristics help boost their self-confidence and courage to move to the next opportunity. Seth Godin, entrepreneur and best-selling author, argues in his book The Dip that "winners" have the ability to quit fast, often, and refocus their efforts on something new. "They quit the wrong stuff, stick with the right stuff, and have the guts to do one or the other." The idea is not to give up simply because reaching the goal is challenging but knowing when to make strategic decisions to reallocate time and effort to pursue better opportunities. As Godin says, "Quitting the projects that don't go anywhere is essential if you want to stick out the right ones."

The speed and scale of current technology disruptions have no historical precedent in human history. From now until 2050, this trend will continue to accelerate. Singular humans know that persistence is key to overcoming obstacles and necessary to live an extraordinary life. They routinely evaluate their decisions, perspectives, and ideas to ensure they align with where the world is heading. If they detect their plans are not the best fit, they are not afraid to find alternative solutions. They know when to change direction and have the courage to embrace new opportunities and challenges. They know the trick to succeeding in a future in flux is to take consistent actions and be willing to change when things no longer align with their purpose and long-term goals.

Future-Ready **Mindset**

Singular humans are forward-thinking individuals. They are supported by their optimistic yet realistic views of the future. They recognize and appreciate the importance of their past as the foundation of who they are as human beings—their core values, beliefs, and integrity have been developed throughout the years and will continue to shape and guide their actions and decisions. However, they do not use their past achievements as the only gauge for future success; instead, they develop strong foresight to embrace today's challenges and opportunities to imagine and anticipate their ideal future. They are not distracted or afraid of the dystopian or negative stories about the future and how technology could destroy society and the human species in the decades ahead. Instead, they focus on developing human skills to quickly respond and adapt to the never-ending transformations and how to be essential in a world that is becoming entirely digital. Their life decisions favor forward-looking outcomes instead of short-term rewards and impulse gratification.

The formula for living a purposeful life is finding the balance between creativity, imagination, realism, and optimism. Singular humans live in the present and have high levels of awareness about emerging trends shaping the world so they can adjust their actions, decisions, and mindset toward the future they want to build in the decades ahead. Singular humans do not orient their lives to wishful thinking; they move forward even if they don't have all questions answered. They leverage the information, resources, and skills available at any given time—combined with their intuition, an advantage over robots, which cannot replicate this true human capability—to pursue their desired outcomes consistently and courageously.

Although the road to 2050 will be similar to a roller coaster, the singular humans view everyday challenges and problems as opportunities to build new possibilities. They see the world differently—they see a more optimistic future and brighter pathways to personal growth. They live their lives to the fullest and don't ignore the things around them. Instead, they embrace the ever-changing circumstances to build a foundation for tomorrow while adjusting their attitude toward a future that inspires them. The present gives them the foundation for personal growth, but the future keeps them curious, passionate and motivated to create a meaningful life. They realize we are just scratching the surface of AI, virtual spaces, longevity, the future of jobs, and the digital economy—and they are preparing for it. Singular humans are bold visionaries who recognize that by imagining the future they want to build, they can identify the actions, resources, and connections they need to make to get there.

Singular humans are futuristic in their thinking and actions; instead of imagining the future with pessimism and concern, they activate a mindset of optimism and creativity and turn anxieties and fears into courage. Their future-ready mindset is an essential resource to prepare them to embrace an unknown future and achieve their grandest dreams.

As the world faces more frequent and severe disruptions and becomes more unpredictable than ever, singular humans favor flexibility, adaptation, and value creation. Of course, goals are vital, but they do not spend energy on inflexible outcomes. Instead, they constantly gauge whether they are on a purposeful path to achieve their human potential and long-term objectives.

Integrity

A tsunami of misinformation, discrimination, divisiveness and harmful content is hitting the internet without any filters to stop it from spreading. As machines and algorithms become more complex and sophisticated, the world will need more ethical minds to develop and use intelligent machines for work, education, and play.

Singular humans understand integrity in the virtual and physical worlds needs to come from the individual level. When they are online, they contribute by policing themselves and their behaviors and avoiding the spread of harmful and misleading content. Bioethicist Larry Churchill states, "Ethics, understood as the capacity to think critically about moral values and direct our actions in terms of such values, is a generic human capacity." Singular humans are honest individuals by nature and adhere to strong ethical principles. This characteristic makes them distinct. They question whether something is appropriate or damaging in any situation. There will be a critical need for more ethical humans to keep systemic issues such as inequality, discrimination, and bias in check throughout the twenty-first century.

The internet, AI, and social media platforms powered by ever-smarter algorithms will continue to expand their reach to unimaginable and uncharted places, opening new dialogues and connections. As these technologies become widely available, people will access a borderless digital environment. Unfortunately, some will use them to harm others instead of enhancing life and humanity's progress. A singular human thinks differently and consciously as they evaluate whether something is right or wrong in any interaction, relationship, and decision. They strive for social inclusiveness and use technology to connect brilliant minds rather than separate them. Regardless of gender, politics, or religious beliefs, they think collectively and focus on overall human well-being. Individuals who can make fair judgments and ethical decisions based on careful examination of facts will be invaluable in the coming years. As they grow up, young people need to acquire ethical principles as part of their primary education. Being respectful, trustful, responsible, honest, and authentic will differentiate them from the masses, especially as the world fills with people doing whatever other people—and machines—tell them to do.

A 2050 Scenario:
Past, Present, and Future Generations

Maria has always been a woman ahead of her generation. When only a few young women graduated from college, she pushed herself to earn a degree. In her mid-thirties, Maria started her first job as a secretary. Although her work experience was primarily repetitive, manual, and monotonous, she has always felt motivated and empowered by the opportunity to develop herself. Last century, skills acquired in school or college were sufficient to ensure employment stability throughout many decades, mainly because changes in technology and business were slow and less disruptive compared to the twenty-first century. Maria has progressed her career linearly, operated steadily, and followed step-by-step approaches at work. Most of the time, the same answers or solutions would solve many different problems. Now, at 120 years old, she is proud of her career achievements and, more importantly, for having a solid work ethic, loyalty, and thriftiness.

Fast-forwarding to 2050, the old ways of working and learning are no longer suitable as AI, machine learning, cybersecurity, quantum, cloud computing, and virtual reality have grown exponentially, making skills obsolete faster than ever. These technologies require workers to become more versatile and flexible to adapt to a fast-paced world.

Noah is a forward-thinking person who has always been motivated by turning adversity into opportunity. The disruptions around him were not hard to predict, and with a little bit of oversight and awareness, he has gotten comfortable with the uncomfortable. Self-confidence was critical for him to navigate the radical waves of change successfully. He constantly imagined his life in 2050 and challenged himself to be a change agent for his life and family. At sixty years old, he had navigated financial instability and job displacement since 2021, when he was thirty. What made him ready for the future? Focusing on improving human skills, exercising a future-ready mindset, and developing transferable abilities to help him jump from one opportunity to another. This approach made a lot of people successful in the last three decades, including the first trillionaires of the twenty-first century, who leveraged the tsunami of capital to launch digital-native startups and develop new technologies to improve people's quality of life. Their innovations helped alleviate the climate crisis in many parts of the world, enhance mobility in city centers, accelerate the switch to electric vehicles, reduce the digital divide using satellite internet, and expand access to online education to a larger number of people.

Noah has always focused on gradual and constant improvements to achieve long-term goals and dreams. In 2021, he devised a ten thousand days plan—counting the number of days from that year until 2050. Living from paycheck to paycheck, he wanted to achieve financial freedom. In 2021, people were worried about robots taking over their jobs, and as life expectancy continued to increase, many workers were expected to work longer. As a result, he consistently acquired new skills and saved early to prepare for the future. During those ten thousand days, his curiosity and commitment to learning enabled his skill set and wealth to grow exponentially. Although money came from his efforts, his commitment to self-development and long-term dreams paid him the highest dividends: a financially free and purposeful life.

When AI and robots displaced most people, he didn't wrestle with it. Instead, he harnessed the benefits of technology to serve his needs, incrementally improving his skills so he wouldn't be left behind. Noah has recognized that personal growth is a continuous process, not a destination. In a future that has evolved to be more fluid than ever, expanding his abilities by constantly evaluating and recycling his perspectives, behaviors, attitudes, and assumptions proved vital for him to flourish during the radical changes along the road to 2050.

Eva has learned that adaptability is the secret of any traveler on a long road. From an early age, she has been exploring new possibilities, refining her ideas, failing, and trying again. Technology evolved exponentially, not linearly as in the past. The rapid growth of technologies left many individuals unable to adapt fast enough to cope with the disruptions in their lives and careers. Machines are walking, talking, and behaving like humans, automatically adjusting their characteristics to fit people's moods, emotions, and preferences—some are even having sex with humans. Today, nearly ten billion learners and explorers seamlessly access the internet as devices are closer or even embedded in their bodies. The rapid wave of globalization and digitization has accelerated, and along with it, communication between people has become frictionless.

The exponential growth of internet users has made the number of new entrepreneurs skyrocket. One of them is Eva, who considers herself a knowledge hunter and trendsetter. She has combined her father's lifelong learning spirit and her great-grandmother's integrity and loyalty to help people navigate the fast-growing digital economy. She uses the internet's power to increase people's awareness about technology and prepare individuals, particularly young women, to develop high-demand skills and access more fulfilling jobs. Since the beginning of the century, one thing hasn't changed—people continue to be at the center of humanity's progress. Learning from the first trillionaires' successes, Eva aims to serve people and create cutting-edge innovations to build a more sustainable world. She is already making a lot of strides in the virtual world.

The world is
constantly changing

Anticipation is the key to thriving

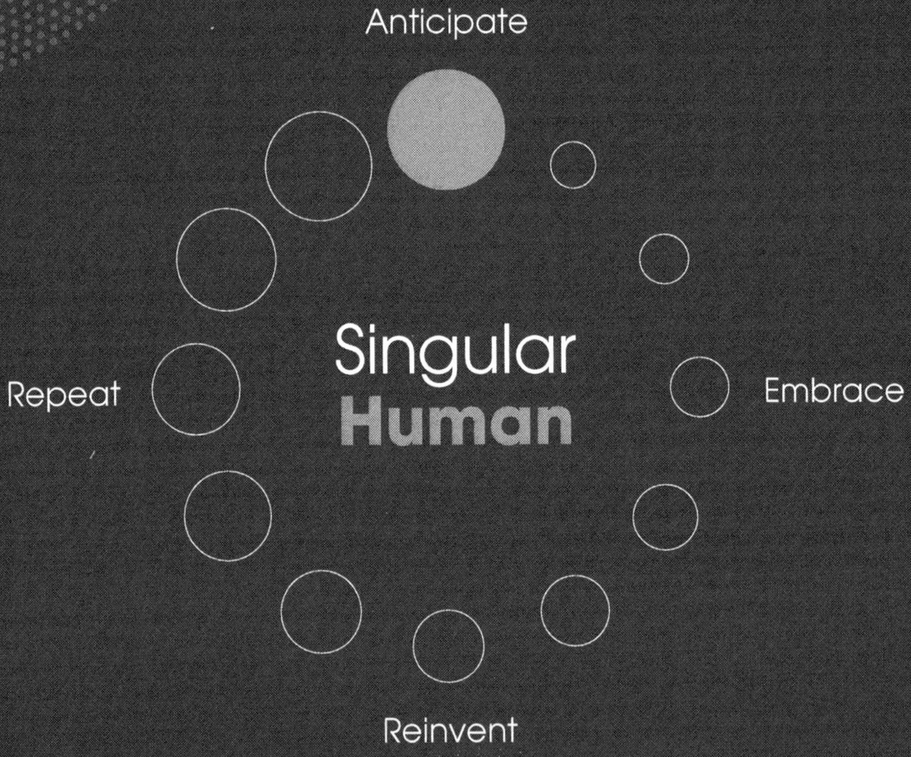

4 Fluidity and Adaptation **for a Successful Future**

Over the coming years, instead of trying to predict how life will unfold, we must reset our thinking and prepare for the unknown by developing essential abilities and equipping ourselves to adapt to unforeseen scenarios in our lives. As we approach 2050, we must assess what is and isn't working, reflecting on successes and failures as an ongoing life ritual. We must pursue continuous improvement and new possibilities in a never-ending loop.

Being paralyzed by the uncertainty of a future that is evolving more unpredictably than ever will likely lead to an unfulfilled life and job displacement. The key is to accept that we can continually improve personally and professionally and constantly remember that our future is a work in progress. True joy comes from appreciating the journey because the destination will keep changing. Daniel Gilbert, psychologist and author of *Stumbling on Happiness,* observed, "Human beings are works in progress that mistakenly think they are finished." He asked people to predict how much they would change over the next decade. He found they greatly underestimated it. According to Gilbert, "Most of us can remember who we were ten years ago, but we find it hard to imagine who we are going to be, and then we mistakenly think that because it's hard to imagine, it's not likely to happen."

Anticipating the **Swells of Life**

The next waves of transformations will be pleasant and smooth for those preparing and equipping themselves with the essential human abilities to adapt to any unforeseen change. Unfortunately, we do not have a swell report that provides future life indicators on a dynamic dashboard, predicting the perfect moment to retire, sell our home, have a baby, or when a robot will take over our job.

In a world where there will be oceans of data flowing from everywhere, making it difficult to digest and synthesize information, we will find ourselves struggling to find the appropriate answers to some life challenges. Decision-making and problem-solving skills will need to be supported by our holistic understanding of the world, combined with our unique human capabilities like intuition, belief, integrity, and lessons from our past successes and failures. Imagining new situations can guide us to recognize the resources we will need to reach our goals and bridge the gap between our present and the future, helping us anticipate the next waves of change and progress toward who we want to become. Imagination, preparation, and adaptation must be a fluid and daily process.

Imagining Your Perfect Future

In The Reach of Imagination, Jacob Bronowski suggests that images can play out events that are not present to our senses and thereby guard the past and create the future. This future may never come to exist. By contrast, a lack of symbolic ideas "cuts off an animal from the past and the future alike and imprisons it in the present. Of all the distinctions between humans and animals, the characteristic that makes us human is the power to work with symbolic images, or the gift of imagination."

Humans have the unique ability to explore concepts outside of their experiences and imagine new scenarios and novel solutions to complex problems. Imagining life thirty years from now and creating a wandering and a future-ready mindset helps us increase our awareness of what it would take to build the future we desire. Imagining ourselves in new situations allows us to identify the resources, skills, and connections required and where we should allocate our energy and attention to create the future we envision. Defining your path forward will transform your imagination into reality. As Albert Einstein said: "Imagination is more important than knowledge. For knowledge is limited to all we know and understand, while imagination embraces the entire world, and all there ever will be to know and understand."

Imagining your life in the future can help you explore many new possibilities and discover the abilities needed to achieve your life goals and dreams, boost creativity, and nurture innovation. Imagination enables you to create images in your mind and gives you hope and optimism to chase a more meaningful future. Leveraging our intuition is essential in our decision-making processes, particularly as we navigate uncharted scenarios. Gary Klein, a renowned cognitive psychologist and the author of *The Power of Intuition,* defines intuition as "the way we translate our experience into action." According to him, our intuition "can help us think ahead of the curve by creating expectations, by connecting the dots, by flagging inconsistencies, or by warning us of problems." Although it's subject to error, especially from historical biases, intuition can serve as a powerful tool to anticipate what is coming next and prepare ourselves for rapidly changing environments. As Klein states, "the world is too complex to think ahead using careful analysis of situations. We have to rely on our intuitions instead."

Soon emerging technologies such as virtual reality and

augmented reality will let us teleport from one place to another, experiencing unimagined scenarios. While this will allow us to boost our creativity, converting our imagined destination into reality will require determination, persistence, and a clear vision of what we want our future to be. We can learn a lesson from elite athletes. They visualize their goals before a competition and see themselves making the fastest time or scoring the winning point. "If we can have those last thoughts be those productive, powerful, strengthening thoughts, then it increases the likelihood of success," says mental performance coach Nicole Detling. "The more scenarios you go through throughout the course of your career, the more ready you are for any situation that happens."

We tend to imagine the worst-case scenarios—struggles and a life we do not want. We often imagine adverse outcomes or worry about the future. We must shift this thought process and believe the changes ahead will bring us positive results. Optimism can lead us to remain confident in the face of adversity and pursue the transformations we want for our lives. Here is how you can use imagination to guide you to your desired goals and positive outcomes:

- **Think** about what you **want** your life to be like in the future.
- **Picture** the desired outcomes in **detail**. Be as specific as possible—the more details you can imagine, the more precise your directions will be.
- **Identify** actions and **roadblocks** to overcome and develop skills to do so.
- **Picture** yourself celebrating your success.
- **Use imagination** as a continuous **habit** until your imagined future is a reality.

Making Sense of the Changes around You

Immense amounts of data surround us. New insights and information come to us from other people's knowledge, our own experiences, and the digital world at an unprecedented speed—eBooks, articles, news, movies, etc. AI and machine learning systems use algorithms to analyze and process data; humans use a more complex process to make sense of the information coming our way. In his book *Thinking, Fast and Slow,* Daniel Kahneman explains that humans have two systems to process information and make decisions:

- **System 1** operates automatically and quickly, with little or no effort and no sense of voluntary control.
- **System 2** allocates attention to the effortful mental activities that demand it, including complex computations.

The future is evolving faster and more unpredictably than ever. Actively perceiving the changes and disruptions in our personal life and career will be vital to help us imagine, plan, and embrace the future with more confidence while taking more calculated risks. In an always-changing world, the more we interpret the world around us, connect with people, seek new skills, and make quick but rational decisions, the more confident we become to move forward to new frontiers in our lives without being afraid of changes. Using our long-term goals and dreams as a compass to orient our actions can help us avoid impulsive, biased, and poor decisions. For example, long-term planning can help us shape the direction of our lives and careers and enable us to achieve financial stability and security. As Kahneman says, "in addition to improving the emotional quality of life, the deliberate avoidance of exposure to short-term outcomes improves the quality of both decisions and outcomes."

Human beings are constantly worrying about the future. We spend a lot of time fretting over situations that could threaten our health, work, family, or personal finances. We are masters of contemplating events that happened in the past or might occur in the future. About 85 percent of our worries never materialize. Worrying too much can impact our well-being—it creates tension, irritability, fatigue, and makes us feel anxious and overwhelmed.

Sir Winston Churchill, who served as prime minister of the United Kingdom, once said, "When I look back on all these worries, I remember the story of the old man who said on his deathbed that he had a lot of trouble in his life, most of which had never happened." Conversely, not worrying can also be dangerous as caution plays an essential role in our lives. However, worries do not need to be destructive or paralyzing; instead of constantly imagining a negative outcome, we must shift our mindset and view the world and new events more favorably. Shifting our mindset motivates us to act and increases our awareness of what is happening around us. As we become more vigilant for signs of trouble, we can bridge the gap between our current state and future aspirations. Kate Sweeny, associate professor of psychology at the University of California, Riverside, is the co-author of *The Surprising Upsides of Worry.* She argues that fretting can encourage behaviors that lead to better health, more success, and greater well-being: "When it comes to worry, that function is pretty clear: It draws our attention to the fact that there's something we maybe should be doing or preparing for or preventing, and it gives us the motivation to do something about that." Self-awareness and managing our concerns and worries can lead to important choices, as worrying prompts us to do something inconvenient yet protective. For example, if you worry about high cholesterol levels, you might improve your diet. If you have an important client meeting, you will spend more time preparing for it. If you worry about retirement, you will start saving early. If you worry about losing your job to AI and automation, you will need to learn how to collaborate with these technologies.

In his commencement speech at Stanford University, Steve Jobs stated: "You can't connect the dots looking forward; you can only connect them looking backward. So, you have to trust that the dots will somehow connect in your future. You have to trust in something—your gut, destiny, life, karma, whatever." Following his advice in the coming years and decades will help us prepare for the future and stand firm against fears of obsolescence, inertia, and failure.

As a fearless human being, you must respond boldly to the unknowns despite anxiety or worries of failure. In a world of rapid and constant transformations, we only have two options: embrace changes or become obsolete. The fearless human being follows a core principle: act now. By avoiding procrastination and perfectionism, they take gradual and daily actions to achieve their goals and dreams. They control their decisions and avoid becoming paralyzed by what might or might not happen in the future. They take risks, experiment, learn and adapt to new scenarios.

Bias for Action

Your personal mission statement starts with your purpose. Clearly defining your goals and how you will achieve them sets your path forward, allowing you to prepare for changes and act accordingly. Planning in a fast-changing and chaotic environment is a challenging task. Still, you need to consistently develop yourself to fill tomorrow's needs instead of playing catch-up to meet the needs of yesterday. Using your creativity and imagination, picture yourself in your desired future; this will help you identify the essential skills and capabilities you must develop and the limitations you must overcome to progress in life.

In a rapidly changing world, you must favor action over inaction and allocate energy and attention to making rapid and conscientious decisions. Having a bias for action can orient you to respond quickly to new scenarios, help you avoid cumbersome decision-making processes, and prioritize paths aligned with your purpose. It can also propel you to overcome paralysis brought on by uncertainty and adversity and build new approaches for achieving your goals regardless of any roadblocks. We tend to assess the potential risks of our actions when new situations occur; to respond to unimaginable challenges, you must change your mindset and ask yourself what the risk of doing nothing is? Inaction is one of the most significant risks to us in a world where what is relevant today may be irrelevant tomorrow. Being afraid to make decisions or take action may cause you to miss opportunities to achieve your desired future, leaving you frustrated and stuck with the status quo.

In an extraordinarily fluid and digital world, machines and robots will be making faster and more precise calculations, decisions, and recommendations than humans can. Maintaining your human distinction is not about chasing perfection; instead, it is about making consistent improvements while learning and adjusting to new circumstances. Challenge yourself to develop a personal "minimum valuable experimentation (MVE)" framework. Like the minimum viable product, which creates a product version with just enough features to be tested. You can explore and experiment with alternatives and acquire new knowledge while practicing MVE every day. The objective is to constantly assess new situations and validate your new ideas and aspirations quickly, aiming to evaluate how your actions and decisions are helping you to achieve your long-term goals. Reflect often, and consider your next steps based on the lessons from your failures and successes.

Make a decision and commit to it with no regrets. Act on what is essential and excites you. Be the author writing your future and turn the page when it's time to explore new situations, knowing that adversity is part of the personal growth process. You will need to build resilience and recover quickly from difficult conditions. So have courage and keep chasing!

Decision-Making during Uncertain Times

In a fast-changing world, the answers to many questions will be unknown. However, you cannot paralyze decisions or actions due to overthinking or procrastination. The world will not slow down and wait for us to make decisions; in the face of rapid technological advances, we must respond quickly to new life situations with the resources, information, and insights at our disposal.

Technology is breaking barriers between countries, cultures, genders, and the physical and virtual universe. There is no one-size-fits-all solution—global decisions may not work at the local level, and vice versa, what is ethical in one culture may not be applicable in others. AI and algorithms are not yet sophisticated enough to spot these nuances. In a hyperconnected world, there are challenging variables: lack of ethics in many AI systems; an explosion of misinformation; increased historical, cultural, and gender biases included in data models; cybersecurity risks; an increasingly diverse and global workforce and consumer pool. These variables add important considerations and complexities to our decisions and desired outcomes in our personal and professional lives. Therefore, we will need to constantly refresh our assumptions, ideas, and attitudes with more empathetic and inclusive approaches. As technology evolves and reshapes all aspects of our lives, our decisions and actions need to consider a broader range of people and groups; otherwise, we risk making unfair, biased, and in some cases irreversible choices in business and life.

AI, machine learning, big data, and social media platforms will be indispensable tools to gather, analyze, and process large amounts of data and information. But our elevated sense of the world and ability to distinguish between what is right and what is not will be even more essential to make proper decisions in a world that is becoming more diversified and unpredictable than ever. These technologies should complement our decision-making process rather than replacing the need for human judgment and intuition—or doing our thinking for us. Human skills and oversight will continue to be critical in the foreseeable future to prevent machines from making harmful and biased decisions and recommendations. Machines' precision and human intuition are not adversaries; instead, they need to augment each other. Machines have the power to perform massive data analytics, identify patterns, and provide more precise predictions, while humans can contextualize insights in ways that machines cannot, as well as connect unrelated ideas to broader scenarios to build a more holistic view of the world.

The best decision-makers in the twenty-first century, in life and business, will be the individuals who balance the precision, consistency, analytics, and scalability of AI systems with their flexibility, intuition, global mindset, and experiences from their own lives and others. In his book *Tech Trends in Practice,* Bernard Marr defines "big data" as an explosion in information generated as digital technology grows. "At the heart of big data is the idea that the more data you have, the easier it is to gain new insights, and even predict what will happen in the future." However, in a future where the world will evolve unpredictably, harvesting massive amounts of information may not be the most efficient approach to draw conclusions or make predictions. Singular decision-makers recognize that pertinent data and proper synthesis are better than accessing and retaining a sheer volume of information. The key to making more competent and more thoughtful decisions is to be aware of the trends transforming the world and the potential impact on our lives and others. We need to use our imagination, intuition, and experiences and know what we should do with the data and information we have at our disposal.

Problem-Solving during Uncertain Times

One of our most significant advantages over today's machines is that, as humans, we can think through many different contexts and scenarios simultaneously, creating an optimal course of action to address both simple and complex challenges. Tackling issues in an environment in a constant state of flux requires flexibility and a good understanding of the problem's root causes. These skills will provide the confidence and knowledge to recognize acceptable risks and solve challenges that life will frequently bring us in the coming years.

In a future where all the facts and information might not be available, we will need to increase the number of questions in the "five whys" to drill down challenges to their fundamental truths. Our human ability to contextualize knowledge and different scenarios need to come to the fore, primarily by posing pertinent questions at the right time. Trying to propose radical solutions without a basic understanding of the problem often leads to new issues or exacerbates the original one. The uniquely human capacity for creativity, forward-thinking, and ethical judgment will be invaluable. While some people will use AI to help them find information and new solutions to solve challenges, others will lose their thinking and reasoning abilities as they increasingly rely on intelligent machines to solve their problems.

The best problem solvers in the twenty-first century, will be those who can effectively identify and understand the root cause of a challenge they are trying to solve. By breaking down a complex problem into manageable and smaller problems, reexamining its core elements, and combining different ideas, they can find novel and faster solutions. Great problem solvers manage uncertainty and think critically to draw more rational conclusions. Their arguments focus on creative and novel ideas instead of replicating old assumptions for new problems. Solving problems in an ever-changing digital economy will demand awareness of cultural differences to detect and neutralize unconscious biases. The most efficient problem solvers of the twenty-first century find solutions to climate, social, and ethical challenges; they balance financial, social, and environmental outcomes before proposing solutions. As digital technologies and the internet reduce the distances between people by connecting them at an unprecedented scale in virtual environments, decisions and solutions may impact individuals and groups in any corner of the world. The best problem-solvers consider equity, fairness, and inclusiveness to solve critical issues such as discrimination, the digital divide, and inequalities.

Solving problems in the digital era is becoming increasingly complex. As amplified waves of change will force us to respond to new situations and adopt new perspectives quickly, we must ask questions and identify the problem beforehand. Narrowing complex issues into smaller problem statements is a helpful approach to solving novel and complex challenges. Breaking down a complex problem into parts allows you to understand the root causes and take smaller steps towards the solution without becoming paralyzed or overwhelmed by unknowns.

One alternative is adopting Aristotle's first principle to solve complex challenges. The concept was coined more than two thousand years ago by Aristotle, a Greek philosopher, who believed that the best way to understand a subject is to break it down to its most fundamental principles. Many successful business leaders agree with this approach—the key is not working harder but thinking critically, actively questioning every previous assumption about a given problem or situation, then creating new solutions from scratch. Using the first principle is a technique that can guide us to develop new solutions through reverse engineering. It encourages us to critically question the assumptions of the challenge and break it down into essential components, then ask more critical questions that will inspire new ideas and ways to reassemble the pieces.

Avoid Becoming a Laggard

Laggards are late to the party. They are the last people to adopt an innovation. They value traditional methods and are highly averse to change or risk. On the other hand, innovators are the first customers to try a new product. They are, by nature, risk-takers, excited by new ideas and techniques.

The next thirty years will present countless new possibilities for us to experiment with new ideas and situations. However, we need to be prepared to change jobs at any moment due to the rapid advances in technology and always-changing business models. The world will evolve rapidly, being laggards—valuing the traditional ways of living, working, learning, and engaging—will slow our progress.

Choose to be an innovator instead of a laggard in your life's journey. Take risks that move you closer to the life you want to live in the future. Keep looking for new opportunities and explore the unknown to avoid becoming trapped in the status quo. Break down your challenges, visions, and ideas into small, manageable chunks, and always focus on making incremental improvements every day. Most importantly, avoid postponing action. Procrastinating before you leap into the next wave of change may strand you in obsolescence and disappointment.

Change Starts with You

Accountability begins with individuals. As a singular human chasing your best version, you cannot wait for others to take the first step on your behalf. You need to dictate the direction, pace, and scope of change you want in your life and remain open to adapting to unprecedented disruptions life will send your way. Accountability is vital to long-term success. The onus is on you to ensure that your attitudes and decisions align with your goals and the results you desire. You must take ownership and carry out the actions required to make your desired future a reality. Accountability helps you eliminate distractions, roadblocks, and unproductive behaviors to keep you focused on chasing your most important life goals and prevent you from hiding from challenging and uncomfortable situations. Accountability means taking full responsibility for your failures and learning how to avoid repeating them in the future.

Reinventing **Yourself**

Your ability to learn, listen, observe, and communicate will make you stand out in a world where humans and machines will increasingly share the same space. Futurist and philosopher Alvin Toffler once wrote, "The illiterate of the 21st century will not be those who cannot read and write, but those who cannot learn, unlearn, and relearn." Having a fixed mindset and strictly relying upon past achievements will not guarantee a successful life and career. Jobs in every industry are evolving radically, and old habits, experiences, skills are swiftly becoming irrelevant. The gap between relevance and obsolescence will grow more expansive every day. To be at the forefront of change, you will need to be prepared to embrace the next waves of change and adapt to unforeseen scenarios. The digital transformation will accelerate in the years ahead, bringing significant shifts to your personal and professional lives. Reinventing yourself and finding new ways of thinking will help you reach your full potential. Your willingness to adapt to continual technological, economic, and social transformations will make you a singular human and steer you toward your life objectives in the twenty-first century.

Retooling, Repurposing **and Reapplying**

In a world of constant flux, recycling old ways of thinking, embracing new possibilities, and learning new skills will better position you to respond to adversities and challenges. Reinventing yourself involves self-evaluation and strong self-awareness. Therefore, you need to continually recycle your abilities and perspectives to endure the never-ending changes, fluctuations, and unpredictable circumstances around you.

As defined in research published by McKinsey, adaptability is the "critical success factor during periods of transformation and systemic change. It allows us to be faster and better at learning, and it orients us toward the opportunities ahead, not just the challenges." Staying rigid and unwilling to recycle your assumptions, attitudes, and opinions is a formula for disappointment, frustration, and irrelevance in a world that is changing fast. The adaptability challenge depends on how quickly and efficiently you can respond to ongoing pressures. You can either accept the status quo and do more of the same or courageously embrace change as an opportunity to develop yourself.

By increasing your general awareness of how technological, social, and financial trends evolve, you can anticipate new scenarios and respond to them before it is too late to change your path forward. If you wait for a specific event or situation to unfold, you will feel pressured and have limited options to decide which way you want to move forward. At this point, you lose the ability to plan and decide for yourself while being forced to move toward places or situations not aligned with your life goals. Constantly evaluating your progress toward the future you want to build allows you to reconsider your current path. Whether your job, relationship, or education is not contributing to your desired future, it is your responsibility to refocus, move out of your comfort zone, and pursue new alternatives to remain on track. In a world defined by discontinuity, you need to be comfortable with discomfort and look at adversity through a new lens, turning uncertain situations into new possibilities.

Adaptability and resilience are among the most critical skills to navigate the technological upheaval between now and 2050. These abilities will shield you from the frequent, sudden, and severe waves of change during the next three decades. The key to progress in a future of unknowns is to revisit and redefine your old assumptions, abilities, and skills often and consistently. As an intentional learner, you will need to master critical thinking to analyze and reconstruct your thoughts, emotions, perspectives, and opinions. Become a mindful person by practicing self-reflection and self-correction. By reflecting on the lessons from the past and analyzing what went right and wrong, you will be able to adjust your decisions, behaviors, and attitudes to build the future you desire.

The difference between those who will struggle to navigate an uncertain future and those who will thrive relies on their capacity to overcome change. The individuals who remain focused on their long-term objectives and quickly respond to life's obstacles will become more versatile human beings. Your ability to continually reinvent yourself will determine how much you achieve. To maximize your chances of succeeding, you need to keep:

- **Reinventing:** remain flexible, embrace challenges and changes, and adapt to new scenarios.
- **Retooling:** seek continuous learning and self-development.
- **Repurposing:** refocus attention, energy, actions, and thoughts on activities that contribute to your desired future.
- **Reapplying:** successes, wisdom, and experiences, knowing that it may not be the answer for the future even if something worked in the past.

To navigate an unknowable future, exploring and acquiring new knowledge and experiences—learning from your and others' successes and failures—and reinventing yourself will help you build a strong foundation when adversity, discomfort, and difficulties arise. As David Epstein states in his book *Range*, "facing uncertain environments and wicked problems, breadth of experience is invaluable."

Self-reflection and self-assessment will be critical to understanding your strengths and recognizing and buttressing your vulnerabilities in the coming years. Exercising these skills needs to be a continuous process embedded in your everyday activities and habits. Reinvention, repurposing, retooling, and reapplying must happen deliberately, progressively, continually, and in loops. Everything you learn, explore, and discover must become insights to educate yourself so you can adjust your behaviors and abilities to fit a new and changing world. Ask yourself these questions often:

1. What can I learn from my previous successes?
2. How can I avoid making the same mistakes in the future?
3. What could I do differently and better next time?
4. Are these actions aligned with my life and long-term goals?
5. What are the new possibilities I can start creating today to live an extraordinary life in the future?
6. How can I add more value to myself and those around me?

You need to frequently identify and reflect on what is working and what is not working in your life and adjust it accordingly. As the world constantly transforms, continually reflecting on your strengths and weaknesses is vital to navigating the rapid changes shaping our future. It's about discovering, learning, reapplying experiences, and adapting to new circumstances. Herminia Ibarra, professor of organizational behavior at London Business School, discusses this in the book *Range:* "We discover the possibilities by doing, by trying new activities, by building new networks, finding new role models."

Let the machines and robots do repetitive, routine, and tedious tasks. Instead, focus on nurturing your human skills, tolerating your own mistakes, and being willing to accept your fears and imperfections. The goal is to learn from them and improve—and try again. Be open and honest with yourself and get uncomfortable with the things that need to change in your life. It will be impossible to navigate the future in stasis in a world defined by change.

A 2050 Scenario: **Change is the only constant!**

In 2020 Eva was born! Noah and his wife, Olivia, didn't know what the future would bring to them. Mixed emotions: excitement, happiness, love, and fear characterized the first months after Eva was born. Concerns about their daughter's future kept them awake many nights like all first-time parents. They worried about household finances, education, relationships, and raising a child in the digital era. As the world went digital, their goal was not to keep Eva away from virtual spaces. Instead, they wanted her to learn how to govern herself in the digital universe. The disruptions were happening around them: AI systems dictating what they should watch, how and when, social media platforms building virtual worlds, and IoT devices tracking their home and biometrics. The world was changing fast, and they had to make quick but conscious decisions to prepare Eva and themselves for the digital future.

Noah and Olivia created a game called Go Human; the goal was to expose Eva to the internet. They were the only three players. As AI and social media grew exponentially, they played the game consistently to track the disruptions driven by these technologies. Awareness and personal safeguards need to be upgraded rapidly and consistently in the digital age. The game created an environment for open dialogue, critique, and planning. The three of them discussed real-life examples and how to prevent harmful content from affecting their lives. The goal was to increase their understanding of AI and social media's benefits and impact on individuals and society. The more they talked and played together, the more prepared they could responsibly embrace the digital world. What did they learn, and how did it help? They anticipated known issues, and instead of preventing Eva from embarking on the digital journey, they turned their concerns into new ways to explore the issue in front of them. Go Human was not a one-off conversation at the dinner table; it was a continuous loop of learning as new technologies disrupted old ones in a matter of months—and in some cases, days. Although Noah and Olivia created the game for Eva, they both benefited from it. Especially Noah, who increased awareness about key trends defining his job and industry, allowing him to anticipate changes.

It's 2030, automation is widespread and human jobs are in decline. Noah, a first-line manager for an automobile manufacturer, was notified his role has become redundant as AI and robots have automated more than half of the production line. New ways of working redefined companies: flatter organizations, less middle management oversight, and automated systems executing most repetitive activities. When he started as a mechanical engineer on the manufacturing floor, automation was already expanding quickly, but his team was still entirely human. Today less than 20 percent of his coworkers are human beings—the others are machines and intelligent robots. It wasn't hard to see how disruptive AI, automation, and digital technologies were rapidly redefining manufacturing. Plus, the transportation sector changed significantly to support net-zero emissions in the twenty-first century. As a forward-thinking person, his commitment to learning helped him anticipate the disruption caused by AI.

For Noah, becoming obsolete didn't mean unemployment; it meant new opportunities to grow in other areas. Passionate about cars, the transition to a new job was seamless. As gasoline-powered vehicles rapidly declined, he anticipated the latest trends and moved from legacy areas to new emerging projects; he is now building flying cars to eliminate traffic in major cities.

The lesson? Regularly updating his abilities using his ten thousand days rule boosted his confidence and ability to anticipate future scenarios. Through self-development, he moved ahead of time. Unfortunately, some of his coworkers were complacent about the transformations around them and didn't take the same proactive steps as Noah; unfortunately, they were left behind.

Welcome to 2040. The center of economic gravity has shifted from West to East, and China has become the world's biggest economy. The world is cashless, and everything is digital; cryptocurrencies, NFTs, central bank digital currencies, and digital payments made physical cash obsolete years ago. Global debt has soared to record highs. Consequently, many countries have faced financial meltdowns in the last few years, adding even more uncertainty to an already unpredictable future. Thanks to his commitment to achieving financial freedom, Noah knew that early preparation would be critical to navigating times of uncertainty. His anticipation paid off in the last years, facing job displacement, inflation, and rising education costs; his financial literacy allowed him to act before pressures built to the point where altering directions would be much more difficult.

And now it's 2050. One hundred years old is the new sixty. Longevity is exploding, driven by a combination of healthier habits, developments in biotechnology, and precision medicine. Health is monitored by sensors and chips outside and inside our bodies, making diseases more predictable, preventable, and curable. The world has never had more centenarians than today. Governments cannot keep up with the increased cost to support an aged population. AI, robots, and chatbots significantly increased their situational awareness and the ability to respond to more complex and unpredictable situations. Super-intelligence algorithms created a new world for machines with enhanced cognitive skills. Noah is sixty today and continues to reinvent himself as retirement keeps slipping forward in time. Changes were constant, new skills, career shifts, and financial concerns. However, those situations served as learning points to prepare for when the next disruption arrived. Today, retirement is not his most important life goal. Instead, he now wants to make it to the twenty-second century by living a healthy and purposeful life.

Machines and robots should not mimic human sentiment and autonomy or replace human-to-human relationships. Instead, they **need to serve as an intelligent vehicle** to take us where we want to go quickly, consistently, and safely.

5 Technology Is a Vehicle **Drive It Safely**

Humans have always worried about how technology might disrupt and transform our lives and society. The invention of the telephone caused skepticism and concerns that it would tear apart the fabric of society by removing the need for people to meet face-to-face. People feared radio and television would brainwash everyone. Some believed the radio would turn people away from reading or having intimate conversations with one another. In the 1980s, the personal computer age arrived, and computerphobia was suddenly everywhere. Charles Rubin of *Personal Computing* advised wary consumers back then and offered tips on how to treat it:

> *If you're trying to use a personal computer or are considering using one, remember: Allow yourself to be a little ignorant for a while. Plan to spend some time learning; give the computer a chance to prove itself before you decide you can't use it; take things a step at a time; make sure you read the documentation carefully; and finally, don't forget that you're in charge, not the computer.*

When the internet became a global network in the 1990s, people were concerned about how globalization and digitization would affect jobs and economies. Today, ethical concerns are mounting over how AI, machine learning, and automation will impact our lives and well-being. As these technologies grow exponentially, they can reduce or even eliminate the need for human involvement in many jobs, expose companies and individuals to cybercrimes, reduce people's control over their lives, and track our sensitive and personal data inappropriately.

In the 1940s, science fiction author Isaac Asimov defined the Three Laws of Robotics that he felt should guide their development:

- A robot may not injure a human being or, through inaction, allow a human being to come to harm.
- A robot must obey the orders given by human beings except where such orders would conflict with the First Law.
- A robot must protect its own existence as long as such protection does not conflict with the First or Second Law.

Asimov's laws have influenced science and technology for decades. However, given the breadth and speed with which technology has evolved in recent years and how it will continue to grow in the future, his guidance is outdated for today's world. For example, the definition of "robot" has rapidly expanded to unimaginable fields, blurring the virtual and physical frontiers. We have vacuum robots, military drones, virtual assistants, and chatbots. Soon there will be avatars that interact with us in virtual spaces, fully autonomous vehicles that will drive and fly us, and nanobots that could be injected into our bodies to repair cells.

AI is becoming commonplace in our lives—even if we don't realize it. AI and machine learning systems constantly suggest what we should see, hear, say, read, and consume. Increasing awareness about how these systems work will aid us in leveraging these technologies to take us in the direction we want to go instead of being influenced or controlled by them. In addition to eliminating anxieties about general machine intelligence or mass unemployment.

Driving It **Safely**

Advancements in technology will not slow down in the coming decades. According to an analysis completed by PricewaterhouseCoopers (PwC), global GDP could be up to 14 percent higher in 2030 because of the accelerating development and adoption of AI—the equivalent of an additional $15.7 trillion, making it the most significant commercial opportunity in today's fast-changing economy.

The growth of AI will drive enormous opportunities for businesses to transform industries worldwide and propel humanity's progress; likewise, the potential will be tremendous for individuals to access more knowledge, personalized education, and expand their connections. But technology needs to evolve following policies and regulations to ensure digital divides, surveillance, and economic inequalities are not exacerbated even further over the following decades. Moreover, the effects of digital technology on our lives and society will largely depend on how it's wielded. We need to use it as a vehicle and drive it safely.

Today, our smartphones and GPS can help us reach the destination we want to arrive at. Tomorrow, autonomous vehicles will do all the driving for us. But regardless of how advanced technology is, we still need to provide directions to our smartphones, GPS, and autonomous vehicles, not the other way around. We don't simply sit in the car and wait until it decides where to take us. We are responsible for providing it with a destination. We need to use digital technologies and social media networks as vehicles to improve our well-being and intelligence. These technologies need to take us from where we are today to the destination we want to reach in the future by helping us grow, learn, and expand our skills. Allowing technology to control your destination may lead you to undesirable places. The way we use digital technologies needs to align with our life's goals instead of contradicting our values or dictating what we learn, see, read, or how we entertain ourselves. We need our values, purpose, and beliefs to dictate our lives, not an algorithm.

Machines and robots don't have their own minds, but they make inaccurate judgments and mistakes. Building a safe and ethical digital environment for all, which can inspire and catalyze transformations instead of becoming detrimental to society, cannot follow a one-size-fits-all approach but rather a combination of efforts and commitments from key stakeholders. The public and private sectors must agree on standards and ethical foundations to oversee how digital technologies are developed. Technology pioneers and developers must prioritize confidentiality, transparency, explainability, and security when developing AI systems and algorithms. These core principles can protect people against malicious, disrespectful, and harmful content and allow users to comprehend and trust the results created by AI algorithms. We can choose how to use digital technologies; ideally, we will use them to uplift ourselves and others. The way we choose to engage, play, and work with AI and social media will define who we are as human beings in the future.

AI will continue to be the catalyst for humanity's progress and improve the quality of our lives. However, precautions need to be taken to ensure humans and machines can collaborate effectively, safely, and responsibly in the coming years:

1. **Increase digital awareness.** The development and adoption of AI, machine learning, and virtual realities will accelerate

in the years ahead. We must be curious and identify how we can leverage the benefits and avoid the harm of these technologies. Being digitally aware can help us understand how to use technology to improve our health, mental fitness, happiness, intelligence, and expertise. We need to stay alert to vicious and malicious effects on our well-being and society via inappropriate usage of digital technologies and social media platforms. Developing digital skills—the ability to use digital and collaborative platforms, process digital information and content, communicate digitally, and analyze data—will prevent people from becoming obsolete and misinformed due to rapid technological advancements.

2. **Optimize your digital life.** Companies are racing to grab and hold our attention. We already spend a lot of time using digital technologies and engaging with the virtual world. As technologies such as AI, IoT, virtual reality, and augmented reality move closer to our bodies, minds, and thoughts, we will need to constantly ask ourselves how we are allocating our energy, effort, and attention in the digital environment. Virtual distractions and temptations cannot influence our lives. AI, media, and social media networks need to spur innovation and broaden opportunities, increase the value of human-to-human experience, and enhance people's overall satisfaction with life.
3. **Reduce bias**. AI algorithms, particularly those used in decision-making and social media platforms, need to be developed without historical, social, or racial discrimination. Once bias creeps in, it can incite hate speech, misinformation, stereotyping, and even discriminatory violence. Sensitive areas such as hiring, criminal justice, credit analysis, and healthcare increasingly use AI systems to assess large amounts of data and streamline decision-making. If the training model contains biased data, outcomes may negatively impact people's lives.
4. **Build machines to augment, not imitate humans.** Technology pioneers, developers, and engineers should not build machines

to counterfeit human's sense of belonging, emotions, or feelings. AI and machine learning need to complement people's abilities and intelligence in areas that demand highly analytical, precise, and powerful computing rather than replace human judgment, deliberation, and reasoning. Machines and robots should not mimic human sentiment and autonomy or replace human-to-human relationships. Instead, they need to serve as an intelligent vehicle to take us where we want to go quickly, ethically, and safely.

5. **Serve as an inspiration.** Technology needs to promote creativity and inspiration, especially for children and teenagers. The internet needs to be an entry point for kids to immerse themselves in discoveries and boost their imagination and creativity in constructive ways instead of a medium to intensify negativity, self-rejection, social comparison, and identity deterioration.

Social Media **Surveillance**

Social media platforms have rapidly evolved from online posts, photos, comments, and messaging exchanges to vital marketing platforms for small and large businesses to reach larger audiences anywhere in the world.

Every action we take online is being carefully tracked by super-intelligent algorithms that aim to maximize our screen time. Tim Kendall, Facebook's former director of monetization, had a role in making the platform as addictive as cigarettes: "We took a page from Big Tobacco's playbook, working to make our offering addictive at the outset." [. . .] "We sought to mine as much attention as humanly possible and turn it into historically unprecedented profits." Kendall stated: "The social media services that I and others have built over the past 15 years have served to tear people apart with alarming speed and intensity. At the very least, we have eroded our collective understanding—at worst, I fear we are pushing ourselves to the brink of a civil war."

Social media platforms have been an instrumental tool to provide social and economic progress in the last few years. They helped people stay connected with friends and family, build meaningful connections with others who have similar interests, and enhance creativity by allowing anyone to share and access information, ideas, and products at unprecedented scale and speed. Businesses, particularly smaller ones, can use social media to reach larger audiences, connect more directly with existing and new customers, and expand their operations to virtually anywhere on the planet. However, some social media networks have been obsessed with capturing more user attention and growing profits from advertising in recent years. The longer a user stays engaged on their platforms, the more efficiently their algorithms can collect and analyze users' data and preferences to create more targeted and customized advertising campaigns, resulting in increased traffic, revenue, and profits for the platforms and their advertisers. According to Tristan Harris, co-founder of the Center for Humane Technology, "all the clicks we've ever made, all the videos we've watched, all the likes, that all gets brought back into building a more and more accurate model. The model, once you have it, you can predict the kinds of things that person does."

As 73 percent of marketers believe social media has been "somewhat effective" or "very effective" for their business, the race to capture our attention and drive up the amount of time spent on networking platforms is likely to continue in the foreseeable future. Therefore, we need to be vigilant and identify the potentially harmful effects of social media networks on our well-being and prevent ourselves from becoming the content sold to advertisers.

Today over 3.8 billion people are using social media worldwide; this number will increase to almost 4.4 billion by

2025. Users average nearly three hours per day on social networks and messaging alone. These statistics represent the power and leverage these platforms can have over users, countries, and economies. Social media algorithms will become more intelligent and precise to attract and increase users. As they find ways to harness our attention, we need to be careful not to fall into their manipulative strategies, which will come from highly personalized advertisements, stories, and feeds. Worldwide sales through social media platforms were nearly $560 billion in 2020, and this trend will only accelerate in the future. Forecasts suggest that social commerce sales will reach around $2.9 trillion by 2026. Dr. Shoshana Zuboff, a social psychologist, philosopher, and professor at Harvard University, states: "this is a new kind of marketplace now. It's a marketplace that trades exclusively in human futures. Just like there are markets that trade in pork belly futures or oil futures. We now have markets that trade in human futures at scale."

As users spend more time on social media networks, the dissemination of fake news and disinformation has also become a systemic issue. According to a study completed by the Massachusetts Institute of Technology (MIT), false information is 70 percent more likely to be shared on Twitter than the truth, and it can reach 1,500 users six times faster. Another important finding is that humans contribute to the spread of incorrect information; the researchers wrote, "false news spreads farther, faster, deeper, and more broadly than the truth because humans, not robots, are more likely to spread it."

Misinformation, extremism, political polarization, manipulative narratives, and social comparisons existed long before social media networks' existence. However, as information spreads freely on the internet, anything can reach global audiences faster than ever. Opinions, photos, videos, and comments gain exposure at unprecedented speed. Social media platforms have the potential for a promising and dangerous future. Defining clear boundaries and drafting responsible regulations will be necessary to avoid catastrophic impacts. Keeping kids and teenagers off social media forever isn't realistic—the key is managing how they consume it. Jaron Lanier, an American computer philosophy writer, states in *The Social Dilemma:* "If we go down the current status quo for, let's say, another 20 years, we probably destroy our civilization through willful ignorance. We probably fail to meet the challenge of climate change. We probably degrade the world's democracies so that they fall into some sort of bizarre autocratic dysfunction. We probably ruin the global economy. We probably won't survive. I really do view it as existential."

Will Algorithms **Dictate Our Lives?**

Advances in AI, virtual reality, augmented reality, 3D imaging, and IoT will further combine our physical and digital realities. These technologies will create virtual environments where we will engage and immerse ourselves in more realistic ways. But for us to enjoy a safer and more ethical digital world, we will need more transparency from companies, effective data privacy regulations, as well as safety nets to protect us against algorithms that hold the potential to influence our thoughts, opinions, and behaviors. As technology insinuates itself further into our lives, Tristan Harris challenges technology companies to bring "ethical design" to their products and ensure they consider ethical frameworks and consequences behind their decisions.

Governments and policymakers will need to play a more prominent role in creating policies for how internet companies operate and manage the content shared on their platforms. For example, the United States government announced they would examine how regulation could better protect the public from harmful content and consider modifying Section 230 of the

Communications Decency Act. This federal law protects internet companies from liability for user-generated content disseminated on their platforms. Reforming this law will require social media platforms to meet a specific set of standards concerning transparency and data protection to qualify for the benefits of Section 230. Congress could also propose carve-outs from Section 230 liability protection by holding social media outlets liable for user-generated disinformation or hateful content.

Realistically, we cannot wait until policies are created or companies change their business strategies. Destructive online behaviors impact communities and our well-being, and this will only worsen if we don't control what we want to see, read, and share online. As an active digital citizen, you cannot allow yourself to become an exploitable resource or simply a product; more importantly, you cannot allow some companies to persuade and manipulate you into anything that doesn't align with your values. You must build solid barriers and monitor your online activity by limiting your exposure to digital platforms, toxic influencers, and online communities that can spread harmful content with unprecedented efficiency, speed, and reach. Protect your identity and reputation, and be aware that posting sensitive information can harm you and others. Your priority is to leverage technology to create positive outcomes instead of letting them dictate your life. As Harris states in *The Social Dilemma:* "It's not about technology being the existential threat. It's the technology's ability to bring out the worst in society, and the worst in society being the existential threat."

Building Healthy **Human-to-Machine Relationships**

Technology has always come with inherent risks, and AI is not more—or less—of a threat. As Ray Kurzweil describes: "Technology has always been a double-edged sword. Fire kept us warm, cooked our food, and burned down our houses." He states that technology can play an important role in augmenting and reinforcing our human abilities instead of making us obsolete. "My view is not that artificial intelligence is going to displace us, it's going to enhance us. It already does."

According to a Gallup survey, nearly 73 percent of adults in the United States believe AI will "eliminate more jobs than it creates," according to a Gallup survey. The same study found that 23 percent of people were "worried" or "very worried" that automation would affect them personally. As the need for human workers in many industries will decrease, developing human and digital skills is essential to adapt to the next waves of automation.

Newsweek published an article in 1997 titled "Man vs. Machine," the first line reads, "The hope of humanity is in a good mood." The article was published a few days before the chess rematch between Garry Kasparov and IBM's supercomputer Deep Blue. The match gained worldwide attention because it was the first defeat of a reigning world chess champion by a computer under tournament conditions. But also because of increased concerns about AI's power over human beings. The article calls the chess rematch "the brain's last stand."

> *Kasparov represents our species in a battle that may one day become the prime landmark in technology's ineluctable march to surpass its makers. [. . .] The question has arisen: can the master of pursuit once thought of as the pinnacle of human intelligence beat something that has never lived? [. . .] What would it really mean if Deep Blue won? Some people have argued whether chess would be diminished by this upheaval. The real question is whether we would be diminished. [. . .] It [the supercomputer] does not think; it crunches numbers and evaluates chess positions. Billions of chess positions. The human brain, on the other hand, looks ahead only a few moves but has something that Deep Blue will never attain: consciousness.*

AI has evolved significantly in many fields throughout the last decades. Still, it has not yet advanced to put humanity at existential risk or replace human labor entirely, mainly as some AI systems are not mature enough to operate autonomously or navigate unfamiliar environments. Instead, they focus on solving specific problems, such as recognizing pictures, interpreting video feeds, spell-checking emails, recommending products, or having basic interactions with humans, for example, responding to simple customer service queries. In addition, machines still lack the abilities and thinking to complete tasks simultaneously like a human.

Alan Turing, considered the father of AI, was the first to test machine learning against a human counterpart with the Turing test. In his 1950 paper, he posed the question, "Can machines think?" He predicted that the Turing test—whether machines could demonstrate intelligent behavior equivalent to or indistinguishable from a human—would be passed fifty years later, or by 2000. A hypothetical moment of AI is the technological singularity, a point in time when machines reach a level of intelligence that exceeds that of humans. Some experts predict that we may reach technological singularity sometime before 2050. Kurzweil claims this is an opportunity for humankind to improve by making us more intelligent: "2029 is the consistent date I have predicted for when an AI will pass a valid Turing test and therefore achieve human levels of intelligence. I have set the date 2045 for the 'Singularity' which is when we will multiply our effective intelligence a billion-fold by merging with the intelligence we have created." Meanwhile, other experts like Elon Musk and physicist Stephen Hawking have expressed concerns that super-intelligent machines could end the human species as we know it.

It remains to be seen whether we will reach a point when machines outsmart human intelligence and threaten our existence, agency, and innate capabilities. Whenever society has faced significant disruptions due to technological advances, humans have continuously developed and adapted with them. Machines have impressive computing power capacity and precision, but our most prominent advantages are thinking creatively, operating autonomously, making mistakes, and learning from them. We can use our intuition and imagination to identify and anticipate the next waves of technological disruptions. These will be precious traits to navigate a world where we will increasingly play, work, and interact with AI.

Dr. Nick Bostrom, a professor at Oxford University and author of *Superintelligence: Paths, Dangers, Strategies,* explains, "There's a mismatch between our level of maturity in terms of our wisdom, our ability to cooperate as a species on the one hand and on the other hand our instrumental ability to use technology to make big changes in the world. It seems like we've grown stronger faster than we've grown wiser."

We live in a time like no other; there are many exciting ways technology, particularly AI and social media platforms, will affect society and the global economy. We shouldn't fear a hypothetical future where super-intelligent machines have the same general learning and abilities as humans. Instead, let's plan how to collaborate with machines to solve critical global issues. During his TED Talk titled "Don't Fear Super-Intelligent AI," Grady Booch, chief scientist for software engineering at IBM Research, states:

> *"We are on an incredible journey of coevolution with our machines. The humans we are today are not the humans we will be then. To worry now about the rise of a superintelligence is in many ways a dangerous distraction because the rise of computing itself brings to us a number of human and societal issues to which we must now attend. How shall we best organize society when the need for human labor diminishes? How can we bring*

understanding and education throughout the globe and still respect our differences? How might we extend and enhance human life through cognitive healthcare?"

Close relationships between humans and machines will define the future of work. But favoring our human relationships needs to remain more valuable than nurturing relationships or building dependency on machines. Human beings are, by nature, a social species that rely on cooperation to survive. Moreover, several distinctive features, including language and culture, suggest that humans mainly depend on social living. We must stay true to our purpose and constantly remember that we all share a responsibility for our collective future. Rabbi Lord Jonathan Sacks notes that the simplest way of safeguarding the future "you" is to strengthen the future "us" in three dimensions: relationships, identities, and responsibilities:

"The only people who will save us from ourselves are we, the people—all of us together. When we move from the politics of me to the politics of "all of us together," we rediscover those beautiful, counterintuitive truths: that a nation is strong when it cares for the weak, that it becomes rich when it cares for the poor, it becomes invulnerable when it cares about the vulnerable. That is what makes great nations."

Building mutual trust, allowing open and respectful communications, and having empathy for others will be critical traits for succeeding in the future. To navigate a world where our energy is increasingly drawn from physical to virtual engagements, we need to strengthen our sense of community, friendship, and cooperation. Meaningful human interactions can be achieved by sympathizing with and accepting diverse perspectives.

A 2050 Scenario:
Navigating the Hyperconnectivity Puzzle

Everything is an entry point for seamless and continuous connectivity. Trillions of interactions occur every second. Machines interact with other machines. Autonomous cars can see and perceive their surroundings and exchange data with other vehicles. Trillions of IoT devices allow buildings, traffic lights, and public transportation in city centers to capture and share information with companies, governments, and citizens. Household robots, appliances, and furniture are connected to the internet. They can track everything in our homes, from monitoring energy and water usage to managing what is inside our refrigerators. Our clothes, toothbrushes, and toilets are always on and tracking our health and fitness conditions, identifying if anything in our body is not functioning correctly. Humans are talking with machines and digital assistants as if they were best friends; some people even treat them as confidants and therapists and are more comfortable sharing secrets with their digital devices than with friends and family. Some people have linked their brains directly to the internet to keep up with the fast-paced advancements of AI. By connecting their minds to the internet, they can seamlessly engage and converse with other people.

The Big Brother feeling of everything being connected and monitored is real! Anything with a screen is going to museums. Information and data have exploded and are flowing freely and effortlessly from everywhere. New technologies have emerged while others have become irrelevant very quickly. Technology has transformed fundamental aspects of human behavior. Some people are taking advantage of it by sending their instructions and directions to computers. Still, others outsource their thinking and decisions to machines, waiting to be told what to do. The latter is being left behind as their cognitive capabilities are being challenged in multiple ways, including their capacity for analytical thinking, memory, focus, creativity, and mental resilience.

But one thing hasn't changed—the need for meaningful human connections, personal agency, and autonomy. Interested to know what happened with the Go Human game created by Eva when she was a little kid? Almost two decades later, it has evolved! A small game that started at home became viral as people from across the globe tried to find solutions to solve the digital pandemic. As some computers reach or surpass the human brain's capacity, the singular humans thrive in the hyperconnected era. Instead of being consumed or influenced by machines, they use the power of technologies to learn, work, and entertain themselves.

In this hyperconnected world, everything is talking: person to person, person to machine, and machine to machine. Humans no longer need a screen or keyboard to interact with machines and the digital universe—AI, machine learning, and robots now fully understand what being, thinking, and acting like a human looks like. To help increase people's awareness and create a more equitable world, Eva passionately expanded Go Human to connect people with meaningful and rewarding information and relationships in the virtual world. In addition, she has been encouraging people to develop healthy habits and monitor their behaviors when using digital technologies.

There has never been a one-size-fits-all approach, and there never will be. Still, the secret to safely and responsibly navigating the hyperconnectivity puzzle of the twenty-first century is consistently increasing digital awareness, fostering open dialogue, and upholding integrity when engaging with intelligent systems. As technology has become pervasive in our daily lives and AI has overtaken many recommendations and decisions, Eva has encouraged people to become aware and protect their vital human traits: purpose, identity, and autonomy. As a result, the individuals following her advice create fruitful digital lives. By immersing themselves in a world of endless possibilities, they feel empowered to improve, advance their lives, meet soul mates, and make a positive difference in the world.

By 2050 we will need more optimistic, audacious, and relentless adults excited to work hard to disrupt and reinvent the world in positive and more fulfilling ways. Teaching today's youth to follow yesterday's rules **will not prepare them to embrace the unknowns** and adapt to the fast-moving future.

6 The Adults of Tomorrow

As the time span between technology generations shrinks faster than ever, disrupting education, business models, and the workplace, the youth need to be prepared to overcome uncertainty and obstacles in a future defined by discontinuity.

By 2050 there will be two kinds of adults: complacent and forward-thinking. The complacent adults, or those unaware of the dangers and rapid advances of new technologies, don't detect warning signs that change is about to occur and don't seek new opportunities to continue growing personally and professionally. They favor instant gratification, exchanging a future benefit to obtain a less rewarding but more immediate one. In contrast, forward-thinking individuals believe they cannot build a fulfilling future if they don't understand what's happening in the present and how to change direction. They are at the forefront of progress, possess strong foresight, and constantly explore new possibilities. They actively track the breakthroughs around them and stay updated on the latest trends, including how emerging technologies, the economy, and workplace disruptions may impact their lives and jobs. They practice reperception every day; they can see, make sense, and become aware of something new in existing information and trends.

Teaching today's youth to follow yesterday's rules will not prepare them to embrace the unknowns and adapt to the fast-moving future. Instead, they need to be empowered to pursue their dreams with confidence, autonomy, and responsibility for their decisions. What should we teach children and adolescents today to aid them in their journey to adulthood so they can excel in the world of 2050?

Growing up now is very different than it was during the twentieth century; the adults of tomorrow—today's young people—are already living in a more fluid, dynamic, and unpredictable environment. In 1950, people were still healing from the arduous years of World War II. Women experienced familial and societal pressures to focus their aspirations on home, husband, and children instead of higher education. Men were expected to be strong, responsible, and assertive; their parents pressured them to excel in school and athletics and attend college. Children walked to school together, doors were seldom locked, and parents watched each other's children as they played games together and spent time outside. Since that era, economic, technological, cultural, and social revolutions have transformed children's behaviors, characteristics, and interpretations of the world around them.

The internet, social media, and digital devices connect today's young people. They have the chance to become the most well-educated and socially and environmentally responsible generation yet, due to effortless access to information and the internet. According to a survey conducted by the Pew Research Center among Americans ages thirteen and older, they are progressive and pro-government; most see the country's growing racial and ethnic diversity as positive. Young adults aspire to make the world a better place, and they can rely on technology to help them achieve it. Teenagers' relationship with technology characterizes this generation: 95 percent have access to a smartphone, and 45 percent are "almost constantly" online. Eighty-four percent of adults under thirty years old use social media sites—YouTube, Instagram, Snapchat, and TikTok are among their favorite online destinations.

Parents need to educate and empower these young digital citizens to use technology effectively and safely, guiding them to approach it as a vehicle to achieve better outcomes instead of building dependency on it. Since everything we do leaves a digital footprint, they must understand both technology's positive and negative impacts. They need to learn how to use technologies and social media responsibly. For example, understanding how these networks monitor their usage, what they do with their data, and how their algorithms try to manipulate them. Increasing awareness of these factors will help young people reflect and choose more appropriate ways to consume and share information online. In addition, they need to imagine, explore, and experience new possibilities technology can offer them today and in the future. Today's youth will be at the forefront of the most profound technological, environmental, social, and economic revolutions that society has ever experienced. They are the leaders of tomorrow and will thus be responsible for developing the most vital technologies, leading nations, and managing companies that will drive humanity's growth.

As young people grow up more socially aware, they have the opportunity to fight inequalities, misinformation, and divisiveness in many places around the world. They have the chance to be the most connected and engaged generation and tackle some of the world's most significant systemic issues. However, short-term distractions, unhealthy influences—from people and machines—an unwillingness to embrace lifelong learning, and a lack of social and environmental awareness could make them the most disconnected generation ever. Over the coming decades, the individuals who remain unaware of critical social, economic, environmental, and technological advances will become irrelevant. To survive the amplified waves of change, they must possess greater creativity, flexibility, communication, critical thinking, stress tolerance, and become intentional learners.

The opportunities for the youth to flourish will be endless and within their reach. As digital natives, they have the potential to frictionlessly and effortlessly embrace fast-paced technological innovations to create new possibilities for their future. However, they will need to develop human traits such as integrity, persistence, relationship, and empathy to do this properly.

There are five main principles young people need to follow to remain confident in the face of frequent disruptions throughout their journey to adulthood and aid them in achieving the life they want:

1. **Dream big and stay committed.** They need to have clear and realistic long-term dreams—these big dreams never die. Their long-term vision will help them remain inspired, motivated, and committed to thriving despite difficult circumstances and adversities. Rather than asking them what they want to be when they grow up, ask about the changes they want to see in the world and how they would implement them as individuals. Ask what would make them happy in the future. Dreams are essential, but dreams without action and responsibility only amount to wishful thinking. Help them to imagine the journey ahead of them. Ask how they will act if something doesn't occur as they anticipated. Make them the owners of their actions and decisions. They need to understand they cannot control all circumstances, but they can influence their opinions, attitudes, aspirations, behaviors, and conclusions. Ask them to teach you about something they enjoy doing. By asking them all these important questions and staying committed to supporting them, they will feel empowered to explore the unknown.
2. **Keep calm and learn.** They will need to cope with more uncertainty

than earlier generations, so they must adapt to changing and challenging scenarios in their personal and professional lives. They need to understand that embracing change is a necessity and can also be fun and rewarding. They have the potential to live a more vibrant life than any previous generation. To overcome the ongoing changes, they will need to manage higher stress levels and increase their emotional intelligence and self-regulation. A way to avoid panic and stagnation is to understand how the present circumstances and the waves of change reshape their future. By embracing a future-ready mindset and ongoing education, they will equip themselves to respond to future challenges. It is impractical to react to a situation when we don't understand it. By learning, increasing their self and social awareness, and developing human skills such as creativity, resilience, storytelling, communications, and adaptability, they will be better equipped to manage higher levels of stress coming from continual disruptions in their lives.

3. **Balance hybrid realities.** They must constantly balance the time spent between physical and virtual universes, knowing when to remove themselves from the digital world and refocus on the present environment. Developing strong self-worth and self-awareness will help them understand that their identity, personality, and characteristics cannot be changed or hidden behind beauty filters or unrealistic lives and wealth in the virtual universe. To flourish in a future defined by change, they will need to be empowered to keep chasing the best version of themselves in the real world—a version created by prioritizing their identity over other people's opinions or perspectives. They need to learn we are all individuals in progress, and the race for a meaningful life is never-ending. They need to favor building relationships with humans over machines. Learning how to distinguish harmful from beneficial outcomes in the digital world is vital for them to cooperate with machines today and in adulthood. In the future, whatever cannot be automated and digitized will be extremely valuable; to have a successful life and career, they need to focus on developing human skills—these

won't be irrelevant in the future.

4. **Build a digital footprint with humans in mind.** They must be aware that actions, reactions, posts, comments, and photos leave a digital record that can be tracked and monitored. Depending on the content, these can be harmful, misleading, or unethical and negatively impact themselves and others. As the adults of tomorrow, they must understand the potential ethical implications of technologies such as AI and social media and the negative impacts of their misuse. As they grow up and begin considering what changes they want to make in the future, transparency, privacy, fairness, and inclusivity need to be part of this vision. Developing integrity will enable them to create and leverage digital technologies to drive sustainable and ethical outcomes. As future leaders, the youth need to understand that helping and improving people's well-being is more important than corporate profits and personal popularity. Regardless of their occupation, as technologies reshape every facet of human life, industries, and skills requirements, everyone will need a certain level of literacy and comfort around technologies and data analytics—they will be pervasive throughout their lives and careers. Although agreeing on a worldwide definition of ethics is not feasible due to cultural, political, religious, and social differences, today's youth must learn the basic ethical rule now: questioning whether something is right or wrong for the collective. Consider the negative impact of any decision and action. Human well-being needs to be at the center of every decision. The individuals who embrace an ethical framework will become more prepared and valuable adults. Collaborating with AI systems will become a requirement in every job and industry. The world will need ethical professionals to ensure machines are appropriately built and used with integrity, security, and objectivity.

5. **Become a global thinker.** Young people will be the glue that brings the world together, and they will unleash unlimited global

progress. By 2050 nearly everyone on the planet will have access to the internet and the virtual world, opening limitless opportunities for education and a more diverse workforce. To succeed in the future of work, they will need to collaborate with diverse teams and individuals located in virtually every corner of the planet. They will need to be humble, respectful, empathetic, and foster collaboration with people from different nationalities, gender, ages, perspectives, religions, and beliefs. The ability to adapt to diverse scenarios and respect other individuals who might have different ways of perceiving the world will prepare the youth to tackle inequality and discrimination. Still, it will equip them to become essential professionals in a global digital economy.

We need to raise a generation of risk-takers, active and courageous human beings who grow up more socially and environmentally engaged and committed to embracing global perspectives and mindsets. Since younger generations will navigate career changes more often than any previous generation, they will need to develop skills that will endure in a fluid future. To maintain their distinctiveness, they need to learn and practice human values that machines can't replicate: generosity, compassion, gratitude, and integrity. These will serve as the foundation for valuable global citizens.

The Key to a **Better Future**

Today's youth will be the grownups of 2050. They hold the key to shaping their lives and fighting the most pressing social issues. As they grow up willing to drive societal change and transform the world into a better place, they will have the internet at their disposal. They are the most tech-exposed and materially endowed generation that has ever existed.

Thanks to the internet, this generation will have access to

foster collective action and mobilize more people toward common goals, pressure organizations and governments to develop more effective strategies to solve digital divides, mitigate climate impacts, and demand political transparency. Klaus Schwab, founder and executive chairman of the World Economic Forum, states: "Transparency, accountability, trust and a focus on stakeholder capitalism will be key to meeting this generation's ambitions and expectations. We must also entrust in them the power to take the lead to create meaningful change."

AI, robotics, virtual reality, cybersecurity, cloud computing, and other emerging technologies will inspire young individuals to build solutions that complement humanity. Instead of focusing on displacement, they will fight for more transparency in the digital world, more effective ways to mitigate the spread of misinformation, and policies to protect citizens from harmful content and misuse of sensitive data.

The youth needs to be the catalyst to effect meaningful and tangible change by embracing the power of community and driving social transformations at the local level, allowing the impacts to reflect globally. According to a McKinsey survey, sixty-six percent of today's youth believe that communities are created by causes and interests rather than economic backgrounds or educational levels. They will need to prioritize collective goals over personal benefits, acknowledging that collaboration is required to solve global problems. They also hold the weapons to fight against climate impacts, including taking control of their consumerism. Based on an estimate by the Global Footprint Network, the current global rates of consumption require the resources of about 1.75 earths. Meaning it now takes the Earth nearly one year and eight months to regenerate what we use in a year. At this level, we risk exhausting the resources that provide us with fresh water, nutritious food, and clean air. Tomorrow's adults need to increase awareness and constantly revisit their everyday habits and choices.

The youth will chase their independence and purpose to control their own lives and contribute to other people's well-being. They will not allow companies or other individuals to dictate their next career move; instead, they will pursue their aspirations and personal growth. This generation will grow up with a more entrepreneurial mindset than previous generations. Primarily as they use the internet and digital worlds to share their abilities, create movements, and expand their network of friends and business connections. Spending years in the same job role or working for decades for the same company will be passé. Younger generations will question the return on their investment of both money and time versus the value and applicability of the skills they might acquire in traditional academic settings. They will pursue more meaningful work and create their jobs and destinies by embracing entrepreneurship.

In the following decades, today's youth will experience accelerated technological advances, shifting demographics, and societal transformations that will impact how they live and work. Financial education must start early to avoid financial hardships and stress during adulthood; unfortunately, less than 17 percent of high school students in the United States take a semester of personal finance, according to Next Gen Personal Finance. And this is not a new phenomenon; a recent Bankrate survey finds that fewer than four in ten people have enough savings to pay for an emergency $1,000 expense in cash. According to another national poll released by Capital One, Americans answered finances are their number one cause of stress (73 percent), ahead of politics (59 percent), work (49 percent), and family (46 percent).

Financial literacy is a lifelong pursuit, and young people must build solid habits early to prepare them to navigate challenges and adversity. It's never too early to start teaching young people about finances. Research by David Whitebread and Sue Bingham of the University of Cambridge shows that many of our financial habits are set by age seven. They suggested that financial behavior — including the ability to plan and delay gratification — is typically formed in the early years of childhood. This generation must find the right balance of what is more important and purposeful for their futures by building safety nets, resisting short-term distractions, and envisioning more rewarding long-term goals.

Today's youth need to focus on what excites and pleases them. Having something to *look forward to* can help them explore and thrive along their journey to a fruitful future. They need to build a sense of belonging and significance, be prepared to respond to the roadblocks they will face, and attach themselves to the reasons to continue looking forward and pursuing their dreams. Likewise, taking care of themselves and *looking after* the people around them and the greater community can help them become more socially responsible in their personal and professional lives.

By 2050 we will need more optimistic, audacious, and relentless adults excited to work hard to disrupt and reinvent the world in positive and more fulfilling ways. We will need visionaries who can translate knowledge and ideas to unforeseen scenarios and are motivated to improve humanity's well-being by building better and more sustainable living, new ways of working, and educational opportunities for themselves, their families, and the world.

The Future **Builders**

The possibilities for the next generation to imagine and explore new solutions for real-world issues and collaborate in complex and virtual environments on a global scale will be limitless. The adults of tomorrow have the opportunity to be recognized as the creators of the most transformative technologies ever invented, the loudest voice against environmental degradation, and advocate for equality, diversity, and inclusion.

Although it's difficult to draw a future profile of today's

youth, they will be defined by unique attributes that will pave the way for humanity's progress and radically transform how humans live, connect, and work—all propelled by accelerated and disruptive advancements in technology. The tech visionaries, the social champions, the entrepreneurs, the money planners, and the climate ambassadors are coming, and they are hungry to drive global change in the decades ahead. By 2050 they will have reshaped communities in ways we have never experienced before. By leveraging an explosion of knowledge, information, social media platforms, and the billions of new users and minds connected to the internet in the years ahead, they will accelerate the creative economy's growth and profoundly rewrite the next chapters of human history.

The Tech Visionaries

The next generation of technology pioneers, developers, scientists, and physicists was born in the digital era, and their lives have been shaped by technology since they were little kids. Their parents were early adopters of the latest tech: household robots, wearable devices for fitness and health, digital assistants, smart home appliances, autonomous cars, and virtual reality glasses. The tech visionaries can't imagine the world functioning without machines, robots, and algorithms. By 2050 they will be the most tech-savvy professionals.

They will grow empowered, energized, and committed to accelerating humanity's growth. The tech visionaries will combine a range of highly specialized expertise with essential human skills such as creativity, curiosity, critical thinking, and integrity. As they grow more conscientious and empathetic, all their decisions will put human beings at the center of technology development. By 2050 their inventions will have enhanced the lives of billions of people on Planet Earth. But as they grow up inspired by the progress in space exploration, they will continue the work initiated by older technology visionaries and speed up the development of the first human colony on the red planet.

Looking at ways to improve people's lives and power global economic growth, the tech visionaries will be responsible for developing the next phase of AI and machine learning. By 2050 they will have developed highly accurate systems that can speak, read, talk, and recognize images, texts, and faces as good as a human. Still, the most innovative and sophisticated AI systems won't be the ones automating 100 percent of lower-level routine and repetitive tasks. Instead, they will be the machines that can help humans solve the most pressing humanitarian issues, such as climate change, inequality, hunger, and affordable education. The tech visionaries will embrace diverse teams and perspectives, especially to avoid biased algorithms. Girls will grow passionate about solving world problems and breaking paradigms in fields that men have dominated—they will want to reduce the Nobel Prize's gender gap with their innovations and contributions to society.

They will leverage quantum computing's power and capacity to solve problems today's computers cannot handle. They will find ways to apply this technology to financial systems and drug development. Disturbed by the number of people affected by heart disease and cancer, the tech visionaries will be the catalyst to accelerate technologies such as genome editing to correct mutations in affected tissues and cells to treat illnesses and help people enjoy longer and healthier lives.

To protect the environment, they will disrupt traditional business models by creating the technology needed to drive leaps in clean energy and reach net zero by 2050. The tech visionaries will grow up with a mission to speed up the research and development of innovations to reduce CO2 emissions worldwide. They will push for widespread electrification of vehicles, from motorcycles to cars to airplanes. As the world prepares to switch to a fully sustainable energy landscape, the tech visionaries will be involved with leading-edge technologies to accelerate the shift, including the most sophisticated wind turbines, solar panels, and energy storage systems. In addition, they will find better ways to grow and distribute food, accelerate agriculture's digital transformation, drive advances in greener architectures, and build technology to combat deforestation.

The Social Champions

The social champions' mission is to promote social cohesion. They will challenge world leaders to deliver on their promises about climate, equality, security, education, healthcare, and financial stability. They will grow up with a sense of belonging and a desire to be part of something bigger than themselves. They will use digital platforms and social media networks to scale their movements globally and reach more people from every part of the world to promote open, honest, and inclusive conversations. They are passionate about creating solutions to real-world challenges and building peer-to-peer relationships to influence and expand their ideas.

They seek a world that welcomes all people regardless of race, ethnicity, gender identity, or sexual orientation. They don't make empty claims; they are knowledgeable about the most pressing issues impacting their communities and the world. They don't oppose governments solely on political party affiliation. Instead, they recognize to demand something from others or make someone accountable, they need to understand the issue first. So they study, engage, and make informed decisions about topics that can impact people's lives in a way that previous generations haven't. And more importantly, they know that if they raise their voices to expose an issue, they need to be ready to collaborate to find new solutions. They recognize that a social challenge is everybody's problem.

The social champions grow up exposed to political divisiveness, polarization, and scandals. As a result, they will develop broader perspectives about politics and social engagement, and they won't be shy about demanding higher levels of transparency, integrity, and accountability from governments. Through digital platforms, they will reinvent politics—some of them will become politicians themselves. Their mission is to empower citizens and work from the bottom up; they will listen and collaborate. They favor transparency, trust, and execution over dishonesty and empty promises. The social champions will connect governments, companies, entrepreneurs, and citizens to consistently collaborate and commit to important topics such as ethics in technology, income and gender inequalities, and initiatives to reach net-zero emissions.

The populations of developing countries will grow faster than those of developed economies in the coming decades, so many of the social champions will come from places such as China, India, and Africa. Many of these adults will have learned lessons from previous generations; some will still live in areas where poverty is prevalent, but these social champions will become more aware and influential with the expansion of the internet into remote and vulnerable locations. They will fight not only for their rights but for the well-being of their peers and future generations. They will be vocal about sexual harassment, age discrimination, and gender gaps in the workplace. They will instigate action to reduce the digital divide, ensuring more people can develop digital skills and access jobs in high demand in the future.

The Climate Ambassadors

The climate ambassadors will be the adults who grow up motivated to fight the climate emergency and the biodiversity crisis. They will help increase climate literacy at the individual level and hold decision-makers accountable as they will grow impatient with the bureaucracy and politics around climate change. Climate ambassadors will help dissolve the confusion around environmental, social, and governance policies and metrics. They will make their voices heard by educating and positively influencing larger groups of people.

Some will grow up exposed to increased water scarcity, extreme temperatures, droughts, wildfires, and coastal flooding. By 2050 many of them will be climate migrants due to sudden changes in their local environment. Growing up in such harsh situations will leave them frustrated with the slow progress on core issues, but at the same time, they will be more resilient and able to thrive in an uncertain future. They don't want to be victims, and their confidence will push more changes to protect their future and that of the next generations.

They will be the catalyst for the circular economy at work and home. They will find ways to share, lease, reuse, repair, refurbish, and recycle existing materials and products for as long as possible. They are optimistic and confident that their actions will bring prosperity to communities in many parts of the planet. By 2050, as researchers, innovators, teachers, and agents of change, their contributions, solutions, and partnerships with many countries and companies will help these organizations achieve net-zero emissions.

The Entrepreneurs

They will be the thinkers, innovators, problem solvers, and global achievers of the future. They have problems, challenges, and solutions flowing in their veins and an exceptional ability to imagine, anticipate, feel, and judge changing scenarios. Their ability to interpret the key trends shaping the world will help position them in the right place and time to capture the most promising opportunities. The next generation of entrepreneurs has two things in their favor: passion and the internet.

They are aware that a new class of consumers, buyers, influencers, and partners will emerge as nearly ten billion people will be working, buying, selling, co-creating, playing, and interacting on the internet by the middle of the twenty-first century. The explosion of connectivity and data will provide endless opportunities to expand their businesses virtually. In addition, as crowdsourcing, blockchain, online lending, digital banking, and initial coin offerings—which use cryptocurrencies to raise funding—will democratize access to capital, the next generation of entrepreneurs will leverage these technologies to move their disruptive ideas from planning to creation to execution to sales in record time.

The next generation of entrepreneurs will launch businesses in unimagined fields. Leveraging the exponential growth of AI, machine learning, social media platforms, virtual reality, augmented reality, and cloud computing, they will disrupt existing industries by devising flatter and faster structures that blur the lines between diverse sectors. They will work with tech visionaries, social champions, and climate ambassadors to create companies that will reshape the world and human history. They will launch people into space, create new banking apps to service billions of people who don't have a bank account, and develop highly personalized training programs to prepare nearly ten billion people for life in an unknown future. They will create new food production methods to sustain a growing global population. To tackle a new wave of mental health challenges, they will invent novel products to enhance wellness.

To the entrepreneurs, enjoying the journey will be more rewarding than chasing a paycheck. They are motivated to build their future and spend their time in relevant and inspiring ways. Their compass is oriented to purpose, challenges, adding value to people's lives, and personal growth. They favor autonomy and understand that creating solutions to real-world problems gives them more meaning and a sense of fulfillment. They look forward to the future. They target long-term outcomes; instead of jumping from one job to another to cover short-term frustrations or less fulfilling careers. They create their jobs so they won't need to slim social pension benefits—if these still exist when they get older. They consider themselves value creators—they create the trends instead of following them. Their competitive differentiators are their company's purpose, value-driven agenda, and culture—these traits are hard to copy. They favor innovation and simplicity; these characteristics make their startups unique, fast, and elegant. Their businesses support environmental, social, and governance principles not just because customers will view them as sustainable companies but because people and the environment are at the root of their values.

The entrepreneurs are risk-takers and embrace the challenge of being their own boss. They are curious and favor lifelong learning over traditional education. They will leverage the exponential growth of the internet, digital technologies, social media networks, virtual realities, and gaming as the foundation of their businesses to capture the opportunities coming from virtual and diverse environments across the globe.

The Money Planners

They will be the most financially literate generation ever; by witnessing financial hardships within their families and communities experiencing global economic recessions and meltdowns along the way, they recognize the importance of learning about money early in life. They have an elevated sense of awareness that digital technologies, AI, and the rapid growth of automation will significantly disrupt their lives and jobs. They continually acquire new skills to seamlessly transition to new careers and opportunities if they find a better fit or are displaced by machines. Their mission is to remove financial stress from the equation to spend time and energy on what matters for them—their loved ones!

The money planners embrace financial literacy as the foundation for building financial freedom. They tolerate uncertainty because they started planning their financial independence early. They are data-driven individuals and budget their expenses. They believe immediate gratification or extravagance should not replace long-term benefits; this mindset will shield them from distractions and prepare them to be responsive in times of crisis.

By 2050 physical money and paper checks will go the way of the dinosaurs. Money will become increasingly digital, but the financial planners will remain consistent with their approach: earn, plan, and start investing early. Their mission is to avoid the extreme and unnecessary stress created by living paycheck to paycheck. They embrace digital payments, open banking, cryptocurrency, and NFTs. Some will start making money early by leveraging the virtual universe to sell old toys, old-school smartphones, create compelling content, or even design virtual artifacts in exchange for some NFTs.

They are **Digital Citizens**

Digital technologies are increasingly inseparable from almost every aspect of our everyday activities. According to a report from Wunderman Thompson Intelligence, 76 percent of participants say their daily lives and activities depend on technology. That number rises to 79 percent of Gen Z and 80 percent of millennials.

Already we have instant and ubiquitous access to information, and we are all surrounded by data. Each day, five hundred million tweets are posted, 294 billion emails are sent, sixty-five billion messages are exchanged on WhatsApp, and five billion searches are generated. Technologies such as AI, machine learning, 5G, cloud, IoT, and virtual reality will fuel the data-driven world. As a result, the volume of data flowing in the digital world will grow exponentially in the coming years. According to the International Data Corporation, the world's datasphere will grow to 175 zettabytes in 2025; as an illustration, if we stored all that data on DVDs, we would have a stack that could encircle Earth 222 times.

One in five Americans admits to feeling overloaded by information. In the years ahead, the problem will not be accessing the massive volume of news, media, and data created by people and machines every day; the real issue to overcome will be the lack of quality control for the incoming information. As Yuval Harari stated in his book *21 Lessons for the 21st Century,* "People need the ability to make sense of information, to tell the difference between what is important and what is unimportant, and above all to combine many bits of information into a broad picture of the world." He explains, "If this generation lacks a comprehensive view of the cosmos, the future of life will be decided at random." This generation needs to readily absorb new information and respond to complex and frequent challenges. By developing critical thinking to interpret all the data around them, evaluate factual evidence, and assimilate diverse perspectives, they can apply relevant information to make conscious choices and impactful decisions. As the rapid technological advances and the explosion of data and information will transform the future of education and work, critical thinking will be an increasingly necessary skill to succeed in the fast-growing digital economy. Young people need to be comfortable making logical connections between ideas and get used to solving problems as quickly and effectively as possible.

Digital citizens from birth, young people need to use digital technologies and social media platforms responsibly and ethically. They need to be confident and empowered to explore the digital world safely, develop healthy relationships with technology, protect their reputation online, and manage increasing cybersecurity threats. In addition, becoming digitally responsible requires evaluating sources of information, thinking beyond themselves, and communicating effectively in ambiguous and diverse environments. Every time a new technology intersects with their lives, they need to embrace these five basic digital principles to enjoy the virtual world safely, ethically, and responsibly:

1. Leverage digital technologies to drive humanity's progress and augment human capabilities, intelligence, and well-being. Understand the negative impact of unethical and biased AI systems.
2. Engage with people on and offline the digital world, respect their diverse beliefs and viewpoints, and welcome constructive debates. Be a digital activist to combat cyberbullying and online discrimination.
3. Use technology to make their voice heard by mobilizing the public and private sectors to act on the most urgent global challenges. Use social media's power to scale their movements globally.
4. Determine the accuracy, reliability, and validity of online sources to avoid any possibility of spreading harmful or inaccurate information. Distinguish between legitimate and dubious sources. Research and evaluate the reference before reposting articles and news on the internet.
5. Preserve personal and sensitive information. Be alert to potential cyberattacks and suspicious activities in emails, social media, and messaging apps.

Digital citizenship goes beyond conversations about screen time; it is about having digital literacy, responsibility, and accountability when navigating the digital world. A study from the Stanford History Education Group (SHEG) uncovered worrisome results about this generation's inability to evaluate information that flows across the internet. The research shows that students had difficulty distinguishing advertisements from news articles or identifying where the information came from. "Many people assume that because young people are fluent in social media they are equally perceptive about what they find there," said Professor Sam Wineburg, the lead author of the report and founder of SHEG. "Our work shows the opposite."

The youth need to be active citizens who see possibilities instead of problems and opportunities instead of risks. In an era of viral news and posts, ubiquitous and uncredited information, and free-floating opinions, it is more critical than ever that young people develop the necessary understanding to assess and contextualize the media they consume. Jennifer Neda John, a researcher at Stanford University, says, "young people are more likely to believe and pass on misinformation if they feel a sense of common identity with the person who shared it in the first place." According to a survey from Common Sense Media, 60 percent of teenagers who use YouTube to follow current events turn to influencers rather than news organizations.

Solid digital skills have never been more critical. The youth must continuously develop those skills to access high-quality work as nine out of ten jobs will require digital skills in the next decade. As digital technologies and AI permeate every aspect of people's lives and careers, everyone will need to understand those systems' concepts, contexts, and capabilities. As the youth transition to the job market, making sense of data and statistics, having a broader understanding of cybersecurity, and fostering digital ethics will help them become an essential resource in a fast-changing workplace.

Young people need to grow inspired and excited to work with technologies as they will be the individuals developing, using, and selling them in the not-too-distant future. In addition, they need to learn and exercise ethical principles, so today's concerns about biased and unethical AI behaviors are eliminated instead of exacerbated in the future.

The Recipe Young People **Need to Succeed**

We should not underestimate youth's potential by ignoring their creativity, aspirations, recommendations, and actions to challenge long-established worldviews. We need to raise explorers by nature, people who live their lives purposely, with clear direction, and always chasing the best version of themselves. From now until 2050, the adults of tomorrow will need to embrace two key ingredients to succeed in a competitive, digitally-enabled and always-changing world: an explorer mindset and accountability. We need children and adolescents to become obsessed with the idea that the more they explore, discover, and experiment, the more knowledge they will acquire throughout their childhood and adulthood, which will help them build their future and a more purposeful life. Younger generations must remind themselves the future is theirs to imagine and create. To thrive in the rapidly changing workplace, they must focus on integrity and embrace transparency in all relationships and decisions.

We need to encourage young individuals to explore new possibilities outside their comfort zone and help them evaluate progress toward their aspirations and dreams. They need to believe in themselves and understand that their commitment to ongoing education can change their lives for the better. Young people must feel empowered to explore the world and discover how they can contribute to their future, community, and society. The secret to ensuring they are chasing the best versions of themselves is to encourage them to keep discovering new alternatives via trial and error; they need to try, fail, learn, and reapply what they learned in real-world situations.

Younger generations will not have a defined path or all the answers to guide them into the future, and neither will their teachers and parents. They need to be encouraged to use their creativity, imagination, and critical thinking to build a meaningful future and the life they want in the long term. The goal is to flourish in a new world that will challenge them with frequent social, economic, environmental, and technological disruptions. They can either remain idle—join the status quo and watch life's endless possibilities pass by them—or develop their abilities to embrace the unknown. If they can dream it, they can build it, and they will do so by putting their imagination to work and building a desired and honorable future.

They must be resourceful, intelligently using the surrounding environment, connections, and available tools to create something new that adds value to the world. Although no one knows precisely how the world will function in 2050, developing an explorer mindset and looking for alternatives will help them discover unimagined and versatile solutions for global problems. New perspectives and viewpoints will allow them to solve those problems and adapt to unique circumstances.

They will work side by side with robots in the future. Still, to do the things that are more creative and strategic, they will need to focus on building essential human skills that cannot be commodified or made obsolete due to the rapid advances in AI and automation. Balancing digital literacy with human skills such as communication, collaboration, flexibility, conflict resolution, and integrity will be crucial to ensure today's youth are prepared to respond to unfamiliar situations and survive the disruptions of the next decades. A continual and adaptive loop of new experiences, discoveries, transitions, and ongoing learning will replace the traditional linear growth model—education, job, and then retirement—adopted for centuries. Self-improvement, adaptability, and flexibility are essential to succeed in a world under constant transformation. Young people need to embrace challenges, find bold solutions to unforeseen scenarios and remain accountable for their actions and outcomes.

A 2050 Scenario: **The Education Revolution**

In 2021 kids were still following the same traditional education framework from centuries ago. The same classroom setup and learning methods, which resembled manufacturing production lines from the industrial revolution were still used in the early twenty-first century. Not long ago, one person was standing in the front of the room lecturing while twenty-five other people listened and tried to pay attention. Schools organized students by age; they spent one hour on each major. Standardized tests were applied every other week to check retention and whether students had memorized topics from books. They followed instructions, mandates, and rigid curricula throughout their education.

Unfortunately, this linear, rudimentary, standardized education model failed to transform itself as quickly as technology evolved. As a result, some students were left behind as they couldn't make sense of the world around them and build skills to adapt to the constant waves of change.

Many countries worldwide are experiencing increased levels of personal development and economic growth. These countries have accelerated their investments in education and the future of younger generations by ensuring access to lifelong learning, reducing gender inequalities, and closing the digital divide. Schools have become engagement and experimentation labs where AI complements human teachers. As a passionate teacher, Olivia has helped to revolutionize education in recent decades. Today her top priority is to prepare high school students to succeed in the unpredictable job market of the second half of the twenty-first century. Her mission is to ensure these future workers are equipped with unique human-centric skills. As a teacher in 2050, her responsibilities have transformed drastically from being the single source of information in the classroom to becoming a versatile facilitator and moderator of content. The transition was possible thanks to her AI assistant teacher, whose goal is not to replace Olivia but augment her teaching capabilities. Now she can focus exclusively on preparing and coaching students to become highly adaptable and global citizens. Instead of inundating them with unnecessary or useless information, Olivia's job motivates them to pursue learning paths aligned with their life objectives, motivations, and aptitudes. Her goal is to empower them to experiment with what they have discovered and find solutions to solve real-world issues.

While Olivia focuses on empowering and preparing her students for the future, her AI assistant creates individualized content for each person, alleviating Olivia's workload. The analytical capabilities of her AI assistant are immense; it can monitor students' behavior, emotional state, cognitive and physical performance at the individual level. The massive amount of data collected and analyzed by the AI assistant helps Olivia to create tailored learning plans and guide each student based on their specific needs and level of expertise. AI, big data analytics, cloud computing, and virtual reality have transformed Olivia's responsibilities as a teacher and the lives of her students.

While AI develops individually customized learning experiences, virtual reality creates immersive experiences to enable students to teleport themselves to anywhere in the world. This way, they feel more present and learn history where the events unfolded. Furthermore, immersion in virtual spaces is not only about presence—haptic gloves and clothes create a realistic sense of touch; students can feel pressure, texture, and vibration when they touch virtual objects. Science classes are far more interesting as they can see, touch, and feel their experiments. Virtual reality allows them to share the same virtual space with students from other parts of the world. Learning engagement and collaboration have exploded. School has shifted from lecturing to interacting and exchanging ideas with people from different locations and perspectives in real-time.

The curriculum has been revamped, and it now covers a range of real-world topics. Students learn about AI, digital ethics, climate change, social cohesion, inequalities, and money management. Performance assessment has shifted from memorizing to thinking, creating, and solving. Instead of passive listeners, students have become thinkers, doers, inquirers, and achievers. Micro-credentials and certificate programs have replaced high-stakes tests to assess human skills and behaviors in real-world situations. The two possible outcomes—pass or fail—were substituted by continued learning, where skills and expertise gaps are consistently evaluated and addressed based on life's evolving circumstances. The education revolution has helped develop more audacious, dynamic, and versatile global leaders—something the world has been in desperate need of in the last decades.

Human beings have always chased immortality. Believing longevity is achievable is the first step for becoming a centenarian; the second is continually revisiting and **adjusting the way you live, think, and act.**

7 The 150-Year-Old **Human Being**

Life expectancy is increasing globally. In 1990, global life expectancy was sixty-four years. Today this has risen to seventy-three, and by 2050, that number will be nearly eighty. Centenarians worldwide will be around 3.7 million by 2050, increasing from 573,000 in 2021.

Several factors have contributed to people's longevity over the years. For example, many places worldwide have experienced declining child mortality, better management of infectious diseases, improved sanitation, access to clean water, increased medical knowledge, and higher physical and mental health awareness.

Technological innovations are poised to redefine the healthcare sector from reactive care to a proactive, preventive, and personalized system. For example, AI will increasingly allow the earlier detection of life-threatening diseases, provide more precise and individualized care, and enhance the productivity and efficiency of healthcare delivery. Wearable devices and sensors will increasingly allow people and healthcare professionals to monitor health conditions seamlessly and more precisely. Telemedicine will help to cover the shortfall of physicians, but it will also expand access to quality healthcare to more remote and vulnerable areas.

Tissue engineering will be critical to help reduce organ transplant shortages as it focuses on producing artificial tissue and organ substitutes as permanent solutions to replace or repair the damage.

Extending lifespan makes for a promising industry, and the race to find ways to enable humans to live longer is on. In this race, we can find groups currently taking different approaches to allow humans to live healthier and longer lives. According to an article published on Quartz, the *healthspanners* are more concerned with compression of morbidity; their goal is to provide a better quality of life into old age, guiding people to a death that has not been preceded by years of illness. According to surgeon Tenley Albright, during her time at Harvard Medical School, "we were told that our aim should be to help our patients die young as late as possible." Meanwhile, the *tissue tinkers* are interested in immortality or radically extended life. For example, microbiologist Brian Hanley says that cracking the biological code will allow us to "create humans that can live for 10,000 years." Another group is the *cyborgs*, who believe advances in technology will allow our brains or our bodies—or both—to merge with AI. At that point, we will transcend our biological limitations and move toward physical immortality.

The Race to **Extend Life**

In his book *Homo Deus: A Brief History of Tomorrow,* Yuval Harari says that humankind will eventually be divided into a privileged elite—upgraded superhumans who will enjoy unheard-of abilities and unprecedented creativity—and the majority, who will not be upgraded, will consequently be dominated by both computers and the new superhumans. Harari believes the main products of the twenty-first century will be bodies and brains, and the gap between those who can engineer them and those who do not will increase.

Developments in biotechnology, AI, machine learning, nanotechnology, and genome editing are revolutionizing healthcare and signaling the beginning of a new era that might extend human life indefinitely. In the book, *The Future Is Faster Than You Think,* by Peter Diamandis and Steven Kotler, Ray Kurzweil and longevity expert Aubrey de Grey discuss "longevity escape velocity," or the idea that soon science will extend our life expectancy by a year for every year we live. Kurzweil believes this threshold is about a decade away, while de Grey puts it thirty years out. According to Kurzweil, "as they gain traction in the 2030s, nanobots in the bloodstream will destroy pathogens, remove debris, rid our bodies of clots, clogs, and tumors, correct DNA errors and actually reverse the aging process."

Biotechnology is allowing scientists to sequence and edit the human genome. As referenced by Bernard Marr in his book *Tech Trends in Practice,* "As we have a better understanding of how sequences of DNA (genes) are passed on during that division process, we can have a better idea of the impact it will have on a living organism's ability to cope with injury, allergies, food intolerances, hereditary diseases."

In addition to sequencing the human genome, Marr states, "biotechnology is advancing to the point where it's viable to alter the DNA encoded within a cell." This process creates new possibilities for medicine as editing a living organism's characteristics can potentially be changed. For example, "Children could be made immune to illnesses that their parents are susceptible to, crops can be developed that are resistant to pests and diseases, and medicines could be tailored to individuals according to their own genetic makeup."

One method for genome editing is known as CRISPR-Cas9. Although it's likely to be many years before it can be used routinely for humans, it's currently the most versatile and precise method of genetic manipulation. The technology enables geneticists and medical researchers to precisely edit parts of the genome by removing, adding, or altering sections of the DNA sequence.

Someone is added to the United States organ transplant waiting list every nine minutes, totaling over 107,000 people waiting for replacement organs as of 2021. To solve the transplant shortage, which affects thousands of people every year, biotechnology companies are currently working on 3D-printed organ structures that provide artificial tissue as a permanent solution to replace or repair damaged organs. Martine Rothblatt, the creator of United Therapeutics, focuses on harnessing technology to enhance human longevity. One of her projects is manufacturing organs—hearts, lungs, kidneys—so that when they fail, according to Rothblatt, "we can make new ones using a patient's own cells."

For a long time, aging has been viewed as inevitable and unstoppable. Still, things are starting to change as researchers, scientists, and companies challenge the aging process and even reverse it. According to an article published in MIT Technology Review, David Sinclair, a geneticist at Harvard Medical School, argues that "medicine should view aging not as a natural consequence of growing older but as a condition in and of itself. Old age, in his view, is simply a pathology; like all pathologies, it can be successfully treated. Labeling aging as a disease allows scientists, doctors, and biotechnology companies a greater ability to tackle it rather than treating the diseases that accompany it." Progress has been made to classify, and thus treat, aging as a disease; in its latest version of the International Classification of Diseases (ICD-11), the World Health Organization has implemented an extension code for "aging-related" diseases. The classification allows the scientific community to do more advanced research and eventually develop medicines and treatments designed to battle the aging process. The United States Food and Drug Administration (FDA) has implemented strict guidelines for the conditions under which a drug can be licensed and prescribed. Currently, the FDA does not regulate therapies designed to address aging, but this is beginning to shift. In November 2020, the FDA approved the first treatment for Hutchinson-Gilford Progeria Syndrome. This rare genetic disease causes premature aging and death and debilitating effects on people's lives.

Many investors and technology companies are allocating resources to research aging and age-related diseases. For example, Larry Ellison, the co-founder of Oracle, has donated $370 million to aging studies. Peter Diamandis's mission as the co-founder of Human Longevity Inc. is to tackle the diseases associated with aging-related decline and "change the way medicine is practiced by furthering the shift to a preventive, genomic-based medicine model." California-based Unity Biotechnology is a company whose mission is to "extend human healthspan, the period in one's life unburdened by the disease of aging." California Life Company, or Calico for short, is a research and development biotechnology company founded by Bill Maris—and backed by Google—that aims to slow the aging process and combat associated diseases. Maris thinks people can live "for 500 years."

Human beings have always chased immortality. Thousands of years ago, China's first emperor, Qin Shi Huang, became obsessed with it. Hoping to prolong his life, he consumed mercury sulfide. Ironically, his supposed cure may have helped bring on his death. Sixteenth-century Spanish explorer Juan Ponce de León searched for a fountain of youth.

Advances in technology are enhancing the concept of digital immortality; some futurists believe humans will be able to upload their entire minds to computers and become digitally immortal through whole brain emulation. Theoretically, the brain would be dissociated from the body and thus no longer limited to a biological life span. Hossein Rahnama, an entrepreneur and researcher at Ryerson University in Toronto, believes we will come to embrace the digital afterlife. According to an article in MIT Technology Review, Rahnama describes the augmented eternity idea as creating a digital persona that can interact with people on behalf of someone who has passed away. A digital model is made using a person's digital footprint and history. It can be manifested as a text-based chatbot or a 3D-animated character in a virtual reality environment.

Uploading our minds, memories, and knowledge to a new body, robot, or digital avatar could make our life and work a scattered representation with which anyone could interact at any time and anywhere in the world. For example, a lawyer who charges hundreds of dollars per hour could let clients consult their digital avatar for a lower price. Celebrities, politicians, and other public figures could outsource some of their public interactions to digital versions of themselves. Doctors who need to be at different locations simultaneously could allow their digital avatars to meet with patients.

As the race to increase human longevity continues, we must routinely remind ourselves that longevity needs to align with purpose and meaning. It's not worth becoming a programmable machine just to have a longer life. Instead, extending our lifespan needs to be combined with healthy habits, happiness, clear goals, and positive relationships to be worthwhile. These are the features we need to upgrade in our everyday lives continually.

How to Reach Your **Hundredth Birthday**

Recent research reveals that the maximum human lifespan is 150 years. Gero, a Singaporean biotechnology company, recently concluded that "as we age, our body's ability to recover from shocks and stressors weakens in a gradual and predictable way, eventually petering out completely somewhere between 120 and 150 years." With global life expectancy projected to reach eighty years by 2050, how do we close the gap between that number and the human body's expiration date? What will we be able to accomplish with those bonus years of lifespan? The first 150-year-old human being may have already been born. What if it's you?

Advancements in technology, healthcare, and medicine will offer alternatives for us to extend our lifespan and live healthier. But we need to ensure our everyday habits and behaviors contribute to our well-being, particularly preserving our physical and mental health conditions. The Danish Twin Study found that our genes determine only 20 percent of a person's lifespan—the other 80 percent reflects our lifestyle. So how long do you want to live? How well do you want to age? As forward-thinking individuals, we must remain focused on increasing our lifespan and healthspan and hold our mindset firmly to that goal.

Believing longevity is achievable is the first step toward becoming a centenarian; the second is continually revisiting and adjusting the way you live, think, and act. Peter Diamandis discusses the longevity mindset; according to him, how long you live depends on many factors, including your mindset. "You can will yourself to death or will yourself to a longer healthspan." You need to develop a longevity mindset by positively and firmly believing that your mood, behaviors, and habits—sleep, nutrition, exercise, life-work balance, family time—serve as the foundation for an extraordinary life. Living longer needs to start with your long-term vision; break down the steps to make one hundred the new sixty and remain obsessed with that mission and the actions you need to take every day.

Consider what you need to start doing today to positively impact your health and happiness in the next year and decades. What can we learn from centenarians in Sardinia, Italy, and Okinawa, Japan, to live longer and healthier? These two locations are considered blue zones, where living longer is commonplace. The residents of the blue zones live in different countries. Yet, they have commonalities that lead to longer and healthier lives, including four habits that shape their lifestyles and longevity: they move naturally, have the right outlook, eat wisely, and nurture relationships. As expressed by Dan Buettner, National Geographic Fellow, award-winning journalist, producer, and author, "the world's longest-lived people don't pump iron, run marathons or join gyms. Instead, they live in environments that constantly nudge them into moving without thinking about it. They grow gardens and don't have mechanical conveniences for house and yard work." According to Buettner, most centenarians' diets are based on plants and beans. On average, meat—mostly pork—is eaten only five times per month. Before meals, Okinawans remind themselves to stop eating when their stomachs are 80 percent full. Centenarians in blue zones prioritize their sense of belonging and purpose. Okinawans call this ikigai, or "the reason for which you wake up in the morning."

Successful centenarians in blue zones cherish their families and keep them nearby. Susan Pinker, a developmental psychologist and social science columnist for the Wall Street Journal, conducted research in Sardinia and discovered that the lifestyle of centenarians, particularly their social interactions, is a crucial factor in their longevity: "I quickly discovered by being there that in the blue zone, as people age, and indeed across their lifespans, they are always surrounded by extended family, by friends, by neighbors, the priest, the barkeeper, the grocer. People are always there or dropping by. They are never left to live solitary lives." Julianne Holt-Lunstad, a psychology professor at Brigham Young University, produced similar research that showed people with strong social connections—friends, family, neighbors, or colleagues—are 50 percent less likely to die over a given period than those who have fewer social connections. Next, Holt-Lunstad and her team searched for differences between locations or genders: "Does it vary by country? It does not! Is it stronger for men vs. women? Equally strong!"

The Harvard Study of Adult Development has tracked the lives of more than seven hundred men for nearly eighty years, and the most evident message from this research is that good relationships keep us happier and healthier. Over the years, the study has shown that people who fared the best leaned into relationships—with family, friends, and communities. Dr. Robert J. Waldinger, psychiatrist and the fourth director of the study, explains the three important lessons about relationships identified during the investigation:

> *The first [lesson] is that social connections are really good for us, and that loneliness kills. It turns out that people who are more socially connected to family, to friends, to the community, are happier, they are physically healthier, and they live longer than people who are less well connected. People who are more isolated than they want to be from others find that they are less happy, their health declines earlier in midlife, their brain functioning declines sooner and they live shorter lives than people who are not lonely [. . .] It's the quality of your close relationships that matters. It turns out that living in the midst of conflict is really bad for our health. The people who were the most satisfied in their relationships at age 50 were the healthiest at age 80. And good, close relationships seem to buffer us from some of the slings and arrows of getting old. Our most happily partnered men and women reported, in their 80s, that on the days when they had more physical pain, their mood stayed just as happy. But the people who were in unhappy relationships, on the days when they reported more physical pain, it was magnified by more emotional pain [. . .] Good relationships don't just protect our bodies, they protect our brains. The people who are in relationships where they really feel they can count on the other person in times of need, those people's memories stay sharper longer. And the people in relationships where they feel they really can't count on the other one, those are the people who experience earlier memory decline.*

Building healthy, happy, and strong connections is vital to thrive personally and professionally in the twenty-first century; relationships can also contribute to our overall health and well-being. Yet, unfortunately, many people are experiencing loneliness and social isolation. More than one-third of people aged forty-five and older feel lonely, and nearly one-fourth of adults sixty-five and older are socially isolated. Although it's hard to measure the effects precisely, there is evidence that social isolation increases a person's risk of premature death—possibly more than smoking, obesity, and lack of exercise—and loneliness is associated with higher rates of depression, anxiety, and suicide, according to an article published by the Centers for Disease Control and Prevention.

Digital technologies have connected people from all parts of the world, allowing them to interact with other individuals and groups, become part of broader communities, and have their loved ones "virtually close." As the digital world allows people to work, learn, shop, and date online, collaborative tools and social media platforms need to help us reach unimagined places and expand our network in positive ways instead of building meaningless relationships with them. According to a survey published by the health insurer Cigna, heavy social media users feel lonelier than light users.

As the next waves of change, particularly the exponential growth of technology, unfold far more disruptive and faster than ever, we will need to respond to unpredictable changes quickly and manage stresses and uncertainties to thrive in the years and decades ahead. As stated by Kristalina Georgieva, managing director of the International Monetary Fund, "If I had to identify a theme at the outset of the new decade it would be increasing uncertainty." As the future evolves more complex, cultivating resilience and learning how to manage our stress levels will be crucial to navigating new challenges that will be more frequent and less predictable. Early foresight and anticipation of how the waves of change are evolving and potentially impacting our personal lives and careers can aid us in coping with change and adapting to ever-shifting environments.

According to the United Nations, mental health problems are among the ten leading causes of disability in developed and developing countries; depression is ranked third among global diseases and is projected to rank first in 2030. In the next thirty years, we will experience more change than the progress made in the last 250 years, primarily driven by frequent technological disruptions that will transform every realm of our lives and jobs. These disruptions will test our physical, mental, and financial conditions. Remaining flexible and actively embracing challenges that will keep coming our way allows us to be more responsive to change as technology reshapes the world.

As our next generation of leaders, activists, and politicians, supporting the youth to handle life's stresses and adapt to change successfully needs to be a top priority for governments, schools, and parents. Between 10 and 20 percent of children and adolescents globally suffer from mental health disorders, and suicide is the third leading cause of death among people between fifteen and nineteen years old. Nurturing connections, increasing their awareness of world issues such as climate change, inequalities, discrimination, and improving their financial literacy can help young people enhance their social and emotional well-being as they plan for a future full of change.

Turning stress and uncertainty into new opportunities can help you explore new possibilities to live purposefully and healthily as well as advance your career. In addition, staying optimistic, connecting with others, getting physically active, and seeking professional help if needed will help you maintain good mental health.

As technologies blur the lines between reality and virtuality, we will need to manage our "digital health" in addition to our mental and physical conditions to avoid being overwhelmed by the rapid growth of technology, data, virtual spaces, and misinformation. Gerd Leonhard writes in his book *Technology vs. Humanity:* "data, information, and media grows exponentially, so a huge digital obesity challenge is now looming large. We need to take it seriously and tackle it because the digital strain will be even more crippling than physical obesity."

Pursuing longevity is a lifelong journey. To increase our lifespan—and to live one hundred as the new sixty—we must change our lifestyle and the way we approach aging. As human beings, we are impatient and tend to prefer faster solutions and immediate answers, but this is unrealistic if we want to live longer, healthier, and happier lives. The journey to longevity demands energy, commitment, and consistency to defeat life's adversities and challenges. To celebrate your hundredth birthday, you need to pursue long-term objectives and act on what is under your control: shifting your longevity mindset, revisiting everyday behaviors, nurturing relationships, and protecting your mental health.

Purposeful **Work and Life Planning**

As longevity increases worldwide, people will live longer than ever. By 2050, we will have made incredible advances in healthcare and biotechnology, found the cure for some of the deadliest diseases, and improved the lives of billions of people around the world. Living 150 years or even longer might be within our reach. Imagine if you could reach that threshold. But in a future defined by higher levels of uncertainty and discontinuity, focusing on your purpose-span—the extended time of your life that you spend pursuing meaningful things that make you fulfilled—in addition to lifespan and healthspan is vitally important. By cultivating your purpose-span, you can live more meaningfully, healthily, satisfied, and longer despite adverse conditions and obstacles. To increase your purpose-span, you need to continually find the reasons that make you happy and optimistic about the future. You need to remember why you started constantly. Knowing that as you progress and gain more awareness in an always-changing world, your purpose will need to evolve, shift, and reorient continually.

In a future defined by discontinuity and uncertainty, we will need to reinvent ourselves to adapt to new scenarios frequently. As our lives and everything surrounding us change, our measurement of success needs to remain flexible and consider other elements beyond financial returns. Increasing our sense of belonging and community and ensuring our actions improve our well-being and those around us need to be part of our lives' objectives. As relentless explorers, we need to continue discovering new possibilities that connect us with our values, passions, and happiness.

Meaning and purpose will define the future of work. Employees will increasingly seek trust, social cohesion, personal impact, and a healthy work-life balance. According to a Harvard Business Review survey, nine out of ten employees are willing to trade a percentage of their lifetime earnings for greater meaning at work. Some professionals would be willing to forgo 23 percent of their entire future lifetime earnings for work that is always meaningful.

The exponential growth of AI and machine learning, the adoption of hybrid working models, and changing demographics will radically transform the future workplace. These inevitable transformations will reconfigure how companies manage their human assets. Although we cannot predict what jobs will be created in the next decades, companies will continue to depend on humans' creativity, judgment, critical thinking, and ethics to deliver sustainable growth in the future. Therefore, businesses must focus on people-centric cultures to succeed in a dynamic environment. According to Deloitte's 2021 Global Human Capital Trends survey, 45 percent of respondents say building an organizational culture that celebrates growth, adaptability, and resilience is critical to transforming work.

By personalizing work experiences and creating lasting value for their people, organizations will positively transform the future workplace and gain a competitive advantage. The future of work will be about people, their aspirations, attitudes, and the value they bring to themselves, their organizations, and customers. As Bill McDermott, CEO of ServiceNow, says: "we're in a world now that's all about people. We're in a truly human moment where any connection that doesn't absolutely lean into people will go nowhere."

As you pursue a longer, healthier, and happier life, clearly defining what purpose means to you is critical. Defining purpose is one of the most personal questions as each person sees and approaches life differently. Whatever your definition of purpose is, you must allocate energy and align your actions to it. Keep your goals in front of you and imagine how the coming waves of change could impact your personal and professional lives.

A 2050 Scenario: **How to Live Longer, Healthier, and Happier**

Noah has just turned sixty years old. As he contemplates his mortality, he asks his 120-year-old grandmother, Maria, "What is your secret for longevity?" He wants to learn and replicate her best practices to make healthy life choices and live longer for the sake of his family.

When he poses the question to his grandmother, Noah expects a laundry list of longevity tips. But, instead, she smiles and answers simply, "a sense of purpose and a reason to enjoy life." She explains that eating a healthy diet, frequently exercising, valuing connections, and nurturing a future-ready mindset have guided her life since an early age.

Maria has always been conscious about best spending her valuable time throughout her life. She always planned out how she would spend each hour in her diary to maximize the output from her twenty-four hours every day. She always has prioritized activities that would give her purpose, joy, and long-term benefits. Knowing there are no shortcuts in life. Effectively managing time has served as a vital motivator to keep Maria progressing through life's obstacles. Prioritizing time has also helped her to better identify and manage stress during tough times.

Incredulous, Noah asks himself how such simple techniques can succeed. Technology surrounds him all the time; cameras and sensors are embedded in his body to monitor his biometrics at every movement. These devices stream his health data to his AI doctor, which can assess his health conditions in real-time. Recent readings clarify that his diabetes has progressed; his AI doctor was notified he is not responding to medications. It's only a momentary concern, as highly personalized medicine is rapidly developed and nanobots injected into his body can deliver drugs to hard-to-reach areas, helping accelerate his recovery. Radical advances in biotechnology have helped identify and manage his chronic disease and allow him to live a healthier life. Still, Noah hasn't nurtured healthy habits during the last few years. His poor diet, inconsistent exercise, and struggles to manage daily stress have become problematic. He realizes technology is more accessible and curing more diseases; however, to live longer, healthier, and happier and become a centenarian, he needs to reevaluate his daily decisions and lifestyle, just like his grandmother.

Some people are already experiencing the concept of upgrading health. Eva has already sequenced and reengineered her DNA to avoid inheritable diseases from previous generations, including eliminating her nut allergy. Thanks to profound advances in AI, genome editing, wearable devices, and 3D-printed organs, her generation will defy aging and live healthier lives for much longer than previous generations. Eva wants to mirror her great-grandmother's vitality and happiness. She believes longevity is the new normal, and living 150 years old is achievable. Her goal is to live well into the twenty-second century.

Indeed, biotechnological advancements ended up creating a few superhumans. Some people have uploaded their entire library of knowledge and memories to the cloud, as quickly as saving photographs on Apple's iCloud in 2021. Others are trying to find a suitable robot avatar to transfer their minds and thoughts into an artificial body. Meanwhile, some people boost their lifespan indefinitely and even enhance their beauty with new body parts. But many of them are realizing they also need to work on their purpose-span, as living longer is not the only destination. While a few people have expanded life beyond 150-years, others are still trying to make a living and access basic services, education, health care, and quality housing.

As we share **responsibilities with AI,** we can retain our advantage in critical and strategic areas such as **managing, advising, decision-making, negotiating, and engaging,** while machines and algorithms focus on crunching data, recognizing patterns, and performing repetitive and manual tasks.

8 The Future of **(Human) Work**

While breakthroughs in technology will radically transform the future of work and displace some existing jobs, they will also create novel job titles and career possibilities. Although we cannot predict what the job market will look like in 2050 or what jobs will be available, the exponential growth of AI, automation, and robots has the potential to transform all jobs, industries, and professions in the next years and decades. To thrive in such a dynamic environment, we need to be aware of how these technologies are shaping the jobs of tomorrow so we can anticipate changes and advance our careers. In addition, developing essential human-centric skills will be crucial to respond to the abrupt shifts in the digital age. Professionals will need to acquire new skills and recycle old ones; otherwise, they are at risk of being blindsided by unforeseen disruptions and transformations in the job market.

History shows us that technology is a net creator of jobs instead of a destructive force that makes working humans obsolete. As referenced in Yuval Harari's book *21 Lessons for the 21st Century*, "since the beginning of the Industrial Revolution, for every job lost to a machine at least one new job was created, and the average standard of living has increased dramatically." The World Economic

Forum estimates that by 2025, eighty-five million jobs may be displaced by a shift in labor between humans and machines, while ninety-seven million new roles may emerge as an adaptation to the new division of labor between humans, machines, and algorithms.

From now until 2050, linear and lifelong careers followed by a transition to retirement will give way to frequent shifts to jobs to fulfill new business and customer requirements. In addition, the future of work will be fluid, mainly due to the rapid advances in digital technologies and increased focus on delivering environmental, social, and governance commitments. While we may equate technology, particularly AI and automation, with job losses, we should rethink this concern; with every disruptive new technology, we see an increased demand for skilled professionals to support new business models and customers' needs. For example, the automobile industry is shifting its focus to electric vehicle production to align with environmental initiatives; some legacy occupations in this industry will be displaced or reimagined to fit this trend. The rise of autonomous cars and drones will disrupt human drivers, but they will also create many new occupations in maintenance, remote control, data analysis, and cybersecurity. We may see a decline in the demand for human construction workers as 3D-printed homes increase in popularity. Still, the construction sector will create new jobs for humans to develop, maintain, and operate these systems. AI systems will play a pivotal role in e-commerce in the future. At the same time digital platforms will eliminate jobs in retail, new occupations around digital marketing, user experience, fulfillment, and distribution will be in high demand in the years ahead.

In a fast-changing job market, the key will be to ensure that displaced individuals can quickly equip themselves with the required skills to transition to new fields and industries. Investments in training from governments and the private sector will be critical to avoid human workers becoming obsolete and increasing inequalities even further in an AI-driven economy. The decisions we make today will determine the course of our lives in the next decades. Since we know changes are coming, understanding the forces shaping the future of work allows us to identify gaps and prepare ourselves to embrace new possibilities.

Machines Are the **New Coworkers**

As companies leverage AI, robotics, and automation to enter new markets, expand into existing ones, increase business productivity, reduce human errors, and reduce costs, machines will increasingly work alongside humans. AI and machine learning will radically transform all jobs and industries since more than half of today's jobs could be automated by the end of this decade. However, the scale and speed of automation adoption will vary from sector to sector, and it will also vary over time. According to a PwC analysis, professionals in industries like manufacturing and finance, where algorithms can lead to faster, more accurate, and more efficient data analyses assessments, may experience automation in the shorter term, while the development of driverless vehicles could result in a significant impact for the transportation sector in the longer term.

Healthcare could be less affected due to a greater reliance on social interaction and human skills such as empathy and compassion. AI and automation will play a critical role in healthcare to streamline administrative workflow, enhance patient monitoring, improve medical imaging for screenings, and diagnose diseases more accurately. Still, as digital technologies advance, they will work alongside human doctors and nurses instead of replacing them. In education, the same will happen; AIs will help teachers do their jobs better and more efficiently. By automating administrative activities, teachers can redirect their time toward activities such as instruction, engagement, coaching, and advising to help students solve real-life challenges and build meaningful connections.

A Deloitte's annual global survey found that 73 percent of respondents said their organizations have embarked on a path to intelligent automation. As AI, automation, and digital platforms reshape the workforce of the future, professionals across all occupations and industries will need to develop a greater level of comfort working with machines and robots in the workplace. They need to consider effectively collaborating with intelligent systems and leveraging their speed, accuracy, scalability, and analytical capabilities to handle routine and monotonous tasks. As machines perform mundane activities, humans need to focus on finding creative and authentic solutions to solve complex problems, developing "big picture" strategies, and adding value beyond the automated capabilities of machines.

According to the *Future of Jobs* report from the World Economic Forum, around 43 percent of businesses surveyed will reduce their workforce due to technology integration, and 34 percent plan to expand their workforce for the same reason by 2025. AI, automation, and intelligent machines will increasingly help us amplify our cognitive abilities and free us time to focus on value-added activities in the future of work. The human capability to contextualize, enforce ethical principles, and identify data anomalies can help train datasets and help machines improve further by assisting them in performing a broader range of tasks and explaining their outcomes. As AI, machine learning, and virtual assistants infiltrate every realm of our personal lives and careers, humans and machines collaborating will result in better decisions and outcomes for society than a machine or a human working independently.

We will live and work in the most hyperconnected era in history in the next decades. According to 92 percent of British businesses, digital skills are essential abilities for the modern world of work, a report from the Learning & Work Institute shows. At the same time, three in four (76 percent) businesses say that a lack of digital skills would affect the profitability of their business. We cannot approach the future of work as "humans or machines," the reality will be "humans and machines" working together. AI, robots, and automation will help us manage less attractive, physically demanding, or routine work. This radical change will give us more knowledge and insights to effectively boost our critical thinking and decision-making abilities to perform creative and strategic activities. As work becomes digitally-enabled, digital technologies will complement and augment our human capabilities, not entirely replace human labor.

Machines As **Middle Managers**

Current advances in AI, automation, and collaborative platforms combined with the need for more agile, flatter, and customer-oriented organizations are significantly transforming jobs in middle management. Some managers are even being replaced entirely by intelligent machines, robots, and algorithms. As a result, the leadership landscape is changing in many industries. According to a Boston Consulting Group (BCG) study, traditional management has reached a breaking point; 63 percent of managers do not want to stay in management, and just 9 percent of nonmanagers aspire to become managers.

Companies need to support a culture of innovation, accountability, and agility in the digital age; their operating model needs to be prepared to quickly respond to market changes without bureaucracy. The companies that continue to follow the traditional management structure—command-and-control, one-way communication, cascading of ideas, and matrixed organization—will be disrupted by organizations rooted in trust, accountability, simplicity, and collaboration. Empowerment and speed will be core ingredients for market winners to gain a competitive advantage in an environment of rapid change and growing competition. Decisions need to be delegated to the people on the front line of processes, projects, and customer-facing roles. They need more flexibility, speed, and autonomy to turbocharge decision-making. Future-ready companies define their priorities and strategies by favoring simplicity over internal bottlenecks. To promote a culture of empowerment, trust, and accountability, they know that effectively and consistently communicating their business objectives allow individual contributors and self-managed and inclusive teams to operate in an environment of less ambiguity and confusion.

In the years ahead, developing abilities such as cross-functional management, collaboration, communication, persuasion, self-improvement, and conflict management will be vital for employees to succeed in a flat, autonomous, and high-performing organization. As employees operate more independently, they need to feel comfortable working cross-functionally, expand their networks, and build meaningful relationships with key stakeholders from different levels and organizations. In addition, as people become more empowered to make decisions quickly, they will need to develop higher levels of industry expertise and take calculated risks to stay ahead of the competition.

Based on a study done by Gartner, AI and digital technologies such as virtual personal assistants and chatbots will replace almost 69 percent of the average manager's workload by 2024. Soon there will be two types of middle managers. The co-creators will work in tandem with small and flexible teams. Their core responsibilities will be advising, serving, and enabling others to deliver priorities and results - rather than dictating, mandating, and supervising. They are visionaries and active listeners. They guide and support their teams to explore ideas, implement new solutions, and learn from their successes and failures. While co-creators will oversee coaching and mentoring, the

"Robo-managers" will be responsible for performance oversight and reporting. Machines and software powered by AI and machine learning already perform supervision and performance management of drivers, warehouse workers, call center workers, and housekeepers. This trend will only grow in the years ahead as AIs, robots, dashboards, and communication platforms are used to streamline execution, boost collaboration, analyze customer feedback, and provide employee performance.

Job Description **Reset**

Job descriptions as we know them will soon be passé. New jobs and skills will rapidly emerge to respond to unimagined market disruptions. As new jobs are created to respond to new market dynamics, companies—particularly recruiters— will need to transform their talent identification and hiring processes. A curriculum will not be as relevant as intentionality, adaptability, integrity, and emotional intelligence. These traits will be essential to flourish in an unpredictable future.

AI, robots, automation, cloud computing, and collaboration tools will fundamentally redefine how work gets done and rewrite job descriptions and responsibilities across all sectors of the economy. Job security will come from our ability to learn new skills and a willingness to reinvent ourselves frequently. As unimagined jobs emerge in response to the fast changes created by digital technologies and new consumer behaviors, if we want to gain a competitive advantage, we will need to be flexible, creative, and constantly acquire the skills in higher demand.

Fluidity and dynamism will characterize the future workplace. To stay ahead of the rapid advances of digital technologies and automation, we need to be curious and unafraid of making our jobs obsolete. It's not about putting yourself at risk of losing your job or the company you have created; instead, it's about finding newer and more effective ways to add value for yourself, your organization, and your customers. By making your job obsolete, you can meet shifting demands, anticipate the next waves of technology disruptions, and define how the job needs to get done instead of waiting to be told how to change or transition to a new career. We need to experiment, fail, learn, and repeat to create our jobs and destiny—the job market won't stop changing. This continuous loop of adaptive learning will help us to increase our productivity, create our ladder of personal and professional growth, and find new ways to thrive at work.

The future of work needs to evolve beyond location—remote, in person, or a hybrid environment—it needs to be centered around people. Focusing on people's development includes creating an environment where they feel their actions are appreciated. They have a sense of purpose. They can access continuous learning to expand their careers, have the autonomy to make decisions, and have safeguards to protect their physical and mental health, regardless of where they are working.

Remote and virtual work allows companies to access talent and expertise from any part of the world. Employees can improve their work-life balance by transforming the standard nine-to-five schedule into a more flexible workday. However, not all jobs can be performed remotely. Though there is a growing desire to work from home, work location is defined by the employee's tasks. According to the study "What's next for remote work" conducted by McKinsey, 61 percent of the workforce in the United States cannot work for more than a few hours each week remotely, while 17 percent of the workforce could work remotely partially, between one and three days per week. The remaining 22 percent of employees can work remotely between three and five days a week without affecting productivity; these jobs tend to be concentrated among highly skilled and educated workers in a handful of industries, occupations, and geographies. As the future of work evolves, managers will need to find the right balance between on-site and remote work to promote a culture of trust, accountability, empowerment, and flexibility.

Your career, work experiences, and job satisfaction need to be less associated with a location and more with the prospects and opportunities that will allow you to thrive in the future. Your next jobs need to offer you possibilities for accelerated career growth, a sense of belonging, social impact and position you a step closer to achieving your full potential as a human being. Flexibility will help you manage your time and balance work with your personal life. Still, to harvest long-term benefits, you will need to favor work and projects that can satisfy your unique needs and expectations. Purposeful work improves your productivity and maintains your mental wellness; it also unleashes your full potential to continue growing personally and professionally, whether work occurs remotely or on-site.

Skills **Acceleration**

The lifespan of skills will continue to shrink due to advances in digital technologies that are transforming the workplace at a rapid pace. This trend will accelerate the need for a more flexible, adaptable, and multiskilled workforce. As technologies evolve faster than traditional education and learning programs, skills become obsolete sooner. In the era of accelerated digital disruption, you need to be responsible for shaping your experiences at work by focusing on ongoing education, continually acquiring new skills, unlearning irrelevant abilities, and recycling perspectives and assumptions that don't fit in a fast-changing environment. In the past, professional skills learned during college or in the workplace were relevant for decades. Today the half-life of a learned skill is estimated to be five years—two and a half years for technical skills—meaning a skill learned today will be about half as valuable in five years or less. If you want to stay ahead of this curve, ask yourself these questions often:

- Is my current skill set still in demand in the marketplace?
- How can I improve my marketability for the next two to five years?
- What are the fastest-growing industries of the future?

According to the Chief Learning Officer Magazine, three categories define skills longevity: *Perishable skills,* which have a half-life of less than three years, are specific technical skills updated frequently, like organization-specific policies, tools, and specialized processes or those related to particular vendors, platforms, or programming languages. *Semidurable skills* have a half-life of approximately three to seven years and tend to be frameworks with basic knowledge from which field-specific technologies, processes, and tools arise. They are longer lasting than perishable skills but are still likely to be replaced as the field grows and expands. Finally, *durable skills* have a half-life of more than seven years and include skills like effective problem-solving, communication, and leadership—these are more foundational abilities.

We will face continual disruption and transformation in the future of work. Technology will eliminate many existing jobs while reinventing others. However, new ones will also emerge. To succeed in such a dynamic environment, you will need to become an intentional learner and remain flexible to adapt to new ways of working and novel occupations in growing industries. Therefore, retraining cannot be a one-time effort. Complementing your human-centric skills with digital abilities and emotional intelligence will enable you to stay relevant in a global economy.

Blurred Industry **Boundaries**

The ongoing digital transformation connects people, companies, and customers faster and easier. It allows companies to reduce operational costs and streamline access to investments and human resources. It also empowers new players to enter the marketplace and existing companies to target new markets, geographies, and audiences. Together these trends are reshaping customer behaviors and business strategies and redrawing the boundaries for nearly every sector. As we experience rapid growth of digital-native companies, like financial technology (fintech) and educational technology (edtech) startups, boundaries between industries are being blurred and reimagined due to the competition among large and small companies in multiple sectors.

Companies are now facing greater competition and risk of disruption due to digitalization and globalization, which have eroded traditional barriers to entering the market. In the digital era, every company, regardless of its size, can expand its operations to new areas and places thanks to the internet and virtual platforms. This radical shift pushes companies to compete for market share against a range of new entrants and operate in multiple industries. Thereby, in the last decades, lines that historically separated companies, functions, and even industries have blurred at an unprecedented speed.

Over Amazon's twenty-seven-year history, it has grown from an online bookstore to the world's largest e-commerce and cloud computing corporation. From Macintosh to breakthrough innovations like the iPod and iPhone, Apple has pushed the boundaries of technology; now, it's increasing its presence in healthcare by expanding product capabilities and partnering with leading health centers and research institutions. Chinese Tencent provides services ranging from social media to gaming to finance and enterprise solutions, including digital marketing and cloud services. Companies such as SpaceX and Blue Origin are currently working to accelerate space exploration, a field that used to belong to government agencies. Fintech startups seek to automate the delivery of financial services, improve customer experience, and make money management more efficient. Fintech is disrupting traditional financial and physical banking as AI, blockchain, online banking, chatbots, and cryptocurrencies are becoming commonplace. Likewise, edtech is transforming the education sector by playing a critical role in ensuring that students access on-demand knowledge and personalized training programs to develop skills needed to thrive in the digital age. Instead of textbooks and traditional education frameworks, students can use online platforms for learning anywhere, anytime.

In a not-so-distant future, virtual reality, augmented reality, AI, IoT, robotics, blockchain, and many other emerging technologies will blur the barriers between traditional sectors even further and reinvent every industry. As lines between industries are blurred, professionals and entrepreneurs will need to actively track the latest technology disruptions and understand how to navigate the accelerated digital landscape. By understanding market dynamics and new customer demands, they can get ahead of product research and development and create innovations in high demand. In the digital economy, customers are in the driver's seat. Businesses of all sizes and industries will need to develop new business models, adopt innovative distribution channels to differentiate themselves from existing competitors and new entrants, and move into previously inaccessible markets. As discussed in the article "Competing in a world of sectors without borders" published by Mckinsey, the ongoing digital revolution is "reshaping customer expectations and creating the potential for virtually every sector with a distribution component to have its borders redrawn or redefined, at a more rapid pace than we have previously experienced."

Expanding **Value Creation**

Companies will need to increasingly focus on generating lasting value for the ecosystem they operate. To gain a competitive advantage and maintain sustainable growth, they must put key stakeholders—customers, employees, suppliers, communities, investors—at the center of their decisions.

According to an Accenture survey, 65 percent of employees believe organizations should be responsible for leaving their people "net better off" through work, 74 percent of consumers believe that ethical corporate practices and values are an important reason to choose a brand, and 81 percent of sustainable indices outperformed their peer benchmarks in 2020. Companies will need to focus on their environmental and societal impact in addition to their financial outcomes, particularly by showing how they are improving employees' wellbeing, fostering diversity and inclusion, adopting sustainable practices, and making ethical decisions.

Net Zero Tracker recently found that 667 (33 percent) of the 2,000 largest publicly-traded companies globally by revenue have technically committed to a net-zero strategy. Taking a proactive and consistent approach to ESG policies will be vital in the coming years to create economic benefits, long-term success, and shared value for key stakeholders. In addition, companies following environmental and societal standards will be better positioned to attract customers as they seek more sustainable products and greater social credibility.

As the world goes digital, it is estimated that 70 percent of new value created in the economy over the next decade will come from digitally-enabled business models. As customers demand more agility, competitive prices, personalization, and on-demand products, companies must accelerate their digital transformation to provide more value to customers and satisfy their ever-evolving needs and behaviors. As roughly half of the world's population remains unconnected to the internet, the companies and entrepreneurs building capabilities, culture, skills, and digital infrastructure to scale their business to new places and audiences will get ahead of their competitors.

New Jobs **on the Horizon**

By 2025 employers expect that redundant roles will decline from nearly 16 percent to 9 percent. Emerging professions—the future jobs—will grow from 8 percent to 13.5 percent. In the United States, employment will grow by nearly 8 percent by 2030. Job losses in utilities, manufacturing, and retail will be partly offset by new occupations or existing jobs that will undergo significant transformations in skill requirements. Careers in healthcare and STEM occupations will outgrow the overall projection for employment by the end of this decade, led mainly by positions in data and AI, cloud computing, and engineering. As all jobs will be transformed by technologies, which opportunities will be left for our children and us by 2050?

Across all sectors of the economy, the jobs of tomorrow will experience intense human-machine collaboration rather than competition. AI, automation, cloud computing, and virtual reality will redesign many existing jobs and create others to fill new demands. Although AI and algorithms can transform businesses and automate many human activities, we will continue to be at the forefront of jobs that require dealing with unforeseen scenarios. To thrive in the future workplace, professionals will need to adapt to frequent technological advances instead of resisting them; refusing to adopt new processes and technologies can lead to redundancy and replacement by machines and robots.

From now until 2050, the job market transformations will come mainly from growing automation, increased global competition, rapidly-changing skills requirements, and ever-evolving consumer demands. Actively tracking and understanding the industries leading the economy and transforming the world can help you anticipate and develop essential skills to access the jobs of tomorrow.

Technology Will Lead the Way

STEM occupations are expected to grow by 11 percent over the next decade. As we rely on technology to continue driving humanity's progress and tackling some of the world's most pressing issues—including inequality, crime, healthcare, climate change, and education—jobs in STEM fields will continue to expand to fulfill the demand for more innovation. Technologies such as AI, virtual reality, cloud computing, IoT, blockchain, and cybersecurity will continue to drive future growth across industries and accelerate the demand for new roles and skills in the digital economy.

As AI and machine learning become commonplace in all aspects of our lives, careers, and communities, AI solutions need to be implemented responsibly and ethically. Therefore, new roles such as **algorithm bias auditors** and **AI ethical officers** will emerge to supervise diverse committees to ensure intelligent systems are designed without unconscious biases and, as a result, enable fairness, diversity, and inclusion. Their primary responsibility is to create standardized protocols and methods to ensure ethical principles are adequately translated into the development of AI systems. Furthermore, they will track and help eliminate bias in datasets during and after product development to reduce the chances of human bias negatively impacting AI applications, consumers, and society. Additionally, they will be held accountable to make sure people's data is collected, used, and stored safely and responsibly.

We will also need better data protection and privacy when navigating the digital world. More data will be created as more users access the internet, increasing hackers' chances of malicious activity.

As a result, there will be a greater need for technologies, procedures, and processes to safeguard people against cyberattacks, which are growing in scale and severity. As such, demand for cybersecurity professionals like **cyberattack predictors, cybersecurity analysts, and data lawyers** will rise. The cost of cybercrime continues to increase; it's expected to grow from $6 trillion in 2021 to $10.5 trillion by 2025. According to the annual Global CEO survey by PwC, 71 percent of executives are "extremely concerned" about cybersecurity threats, and almost half (48 percent) of American CEOs surveyed plan to increase their investments in cybersecurity and data privacy. The employment of information security analysts is projected to grow by 33 percent over the next few years. Employees like cyberattack predictors, cybersecurity analysts, and data lawyers would also support the implementation of frameworks and procedures for the usage and portability of data. We already see this as an increasing legal requirement, mainly due to antitrust and data protection regulations like GDPR and the draft European Union AI regulation. These professionals will protect and defend data, devices, infrastructure, and people's sensitive and personal information.

As the digital economy expands to virtual realities, we will see a rise in **extended reality travel agents** and **virtual reality entertainers**. These technologies offer alternative ways for people to experience and virtually travel the world. They will create exciting new models for the leisure and hospitality industries through frictionless digital user experiences, allowing people to have a sense of presence and a positive perception of tourist destinations. They will enable users to travel to unimagined locations from anywhere, at any time, and lower cost since there's no need for physical travel. For those who prefer physical travel, extended reality travel agents will help users "test" a destination before deciding whether they wish to go there. They can also help users experience destinations that are remote, dangerous, or need to be preserved from human contact. As virtual spaces expand into new areas such as education, e-commerce, manufacturing, and gaming, the **chief metaverse officer** will build safer and more responsible digital environments. They will create experiences for people from anywhere in the world to connect, communicate, collaborate, sell, and buy in the virtual universe.

Manufacturing workers have already experienced significant disruptions and transformations over the last decades. As machines take over more routine and repetitive activities, manufacturing workers have the opportunity to learn new skills in more promising areas. As a result, we might see a rise of new jobs in the manufacturing sector like **uncrewed vehicle engineers, commercial spacecraft mechanics, commercial spaceflight instructors, drone data scientists, and intelligent factory supervisors**. For example, a drone data scientist could be responsible for overseeing the collection of drone insight surveillance, safety inspections, and operational analysis for more data-driven decisions or preventive maintenance. The core responsibility of an intelligent factory supervisor would be to identify and deploy advanced technologies to enable the self-optimization of connected assets on a production line. This person can oversee robots, evaluate their performance based on how effectively they can perform predetermined tasks, and share this feedback with robot programmers to recommend areas for improvement.

Health and Personal Care

Employment in healthcare occupations will grow 16 percent in the next ten years. Healthcare support will increase by 23 percent, and personal care and services by 22 percent. Nine of the top twenty fastest-growing jobs are in healthcare or related fields. Because of the rapid advancements of AI, machine learning, IoT, and big data in the healthcare sector, doctors and nurses will act as a liaison between medical innovation and caregiving. Doctors must combine their expertise such as cardiology, radiology, ophthalmology, and pediatrics, with digital skills to make more accurate diagnoses and create more tailored treatments. AI systems will help doctors collect patient data, analyze it, make predictions, recommend interventions, and produce a more personalized diagnosis, while **preventive health advisors** (human doctors of the future) will integrate and apply large amounts of data to improve individuals' health and provide patient-centered care. Preventive health advisors will focus on groundbreaking research and technology-forward treatments like gene editing and complex surgical procedures. They will be at the forefront of patient interactions and communications. **Robotic nurse supervisors** are human nurses who will be increasingly valuable to help develop and train caregiving robots. They will use more intelligent robots to perform repetitive and hazardous activities, examine patients, alleviate their increased workload, burnout, and safety issues. They will also supervise teams of robots who will offer companionship, entertainment through games, and social engagement for the elderly.

The Green Economy

To reach net-zero carbon emissions by 2050, global annual investment in clean energy will need to more than triple by 2030 to around $4 trillion. The need to fight climate change will create millions of jobs in the next few years. The International Labour Organization estimates that 24 million jobs globally could be created by the green economy by as early as 2030. In its "World Energy Transitions Outlook," the International Renewable Energy Agency projects that the renewable energy sector could account for 43 million jobs by 2050. Solar photovoltaics will provide the most jobs by 2050, followed by bioenergy, wind, and hydropower.

The transition to a greener economy will create novel job opportunities. These professionals will be at the forefront of technology and business development to accelerate the transition to renewable energy, speed up the switch to electric vehicles, curtail deforestation, and encourage investment in renewables.

Soon **electric vehicle fleet accelerators**, such as developers, technicians, and engineers, will be in even higher demand to drive the most significant makeover in the automobile industry by boosting the production and efficiency of electric vehicles and the infrastructure to support them including charging stations and battery production. **Renewable energy advocate**s will develop solar and wind technology to accelerate the transition to renewable energy sources. They will also build partnerships between the private and public sectors to promote the adoption of these energy sources. **Renewable job transitioners** will retrain fossil fuel workers with new skills to find job opportunities in growing areas within the energy sector. **Green investment planners** will collaborate with the private and public sectors, universities, and entrepreneurs to ensure financing options are available to boost investment in research and technologies needed to reach net-zero emissions in time. **Green agriculture engineers** will revolutionize agriculture by developing autonomous tractors and robots empowered by AI to automate farming—from seed to harvest. These technologies will allow agriculture to become significantly more efficient and enable farmers to produce more food while consuming fewer natural resources. The green agriculture engineers will help farmers leverage drones to disperse fertilizers and optimize field management more efficiently. **Data scientist farmers** will spend more time interpreting data and trends and less time in the field. Their primary responsibility will be supervising autonomous devices and modernizing traditional agricultural practices and processes. They will use data collected from AI, robots, drones, and satellites to make data-driven insights and increase farming efficiency.

Sustainable cities will need new roles like **smart city engineers** and **urban farmers.** Smart city engineers will develop future cities, fulfilling the increased demand for more sustainable urban centers. Roughly 70 percent of the global population will live in cities by

2050, which means urban centers must have sufficient housing, electricity, water, food, education, transportation, and medical services to support nearly seven billion people. To coordinate such massive efforts, we will need engineers to integrate smart cities so their infrastructure expands sustainably and complies with environmental regulations and citizens' demands for greener spaces. In addition, future engineers will need to increasingly combine their technical expertise with knowledge of digital technologies such as AI, 3D printing, virtual reality, and IoT devices.

Cities will need more efficient and sustainable solutions to support a growing population. Smart city engineers will need to create an urban environment where vertical farming and green spaces coexist with residential and commercial centers. A city's infrastructure will need to integrate communications between devices, people, buildings, and vehicles. More innovative infrastructure will also allow for faster and more integrated delivery systems, as door-to-door delivery will become increasingly automated, using robots and even drones.

People and Culture

New roles will also emerge to support employees as they adapt to the oncoming waves of change, especially coaching and mentoring people to navigate an uncertain future. For example, a **hybrid work integrator** will facilitate a flexible working environment as employees adapt and balance remote and in-person work. Their goal is to improve employees' overall productivity, work experience, and satisfaction. A hybrid work integrator will be critical to create and improve policies, processes, and digital tools to foster an open and collaborative environment for employees, managers, and customers. **Worker's skill champions** will support employees displaced by automation through upskilling and reskilling. Their role is to either help displaced employees relocate to other divisions or help them transition to a position in other fast-growing industries. As companies around the world focus on elevating stakeholder capitalism—serving the interests of all their stakeholders, as opposed to only shareholders—into their business models, a role such as **chief social and environmental officer** could be helpful to ensure companies comply with ESG frameworks to create value and sustainable growth. As more people experience burnout due to prolonged stress, **wellness counselors** provide preventive support. These professionals will be responsible for employee well-being and holding leaders accountable for making progress on mental health initiatives. They will promote an environment where people can talk openly about their vulnerabilities and concerns. They are responsible for designing programs to increase mental health awareness and help people cope with depression, anxiety, stress, and other mental disorders.

The Creative Economy

As more than two-thirds of the value created in the economy will come from digitally-enabled platforms, more content will be produced and distributed virtually, driving limitless opportunities for creators. **Digital content hunters** are experts on language and humor; they master building an authentic online culture. They mentor and finance content creators and small companies to reach niche groups. Some will even help a cohort of celebrities become more relatable. Their mission is to find and incentivize creators who produce the most engaging content to expand their communities and prepare them for the billions of new internet users that will arise in the coming decades. Sociability, empathy, credibility, and creativity are at the heart of **creative builders.** These individuals use digital platforms to share information, ideas, and experiences with their audiences. They create a community by crafting authentic content and building social connections rather than solely promoting products or services. They favor human relationships over fabricated and scripted engagements.

They are talented people who know what content resonates with their audiences. The power of digital platforms and social media networks allows them to freely express themselves and create a space to share their creativity, perspectives, and knowledge. Furthermore, they leverage these global-scale platforms to build loyal relationships with a diverse community that follows them to consume versatile, trustworthy, and engaging content.

Increasing Your **Future Employability**

As a result of the continuous transformations of the job market due to the expansion of digital technologies, AI, and automation, the skills and knowledge most needed and valued in the workplace will keep changing. Therefore, to adapt to the fast-changing job market of the future, we need to develop essential skills that will be valuable over time, remain flexible, and reinvent ourselves frequently, regardless of our industry or job title.

Technology and automation have often benefited economies in many parts of the world. Historically, markets became more efficient, productivity increased while the cost of goods decreased, economies expanded, and new industries were created. In the coming years, technologies will grow exponentially, disrupting economies, industries, and social systems. Even if the probability of systemic mass unemployment is low, we cannot allow ourselves to be complacent in the face of potential threats from AI, automation, and intelligent algorithms.

To increase your employability in the future, you need to develop new skills and reinvent yourself continuously. Therefore, ongoing learning will be vital to adapt to the constant transformations in the job market and increase your chances of accessing higher-paid and fulfilling jobs. Moreover, education must become a habit as you share the workplace with AI and machines that are becoming smarter and faster every day.

By 2030 up to 40 percent of all workers in developed countries may need to move into new occupations or upgrade their skill sets significantly. Meaningful and well-paid work opportunities will demand higher levels of expertise, likely favoring more educated and skilled professionals. Jobs that require specialization in a narrow range of routine activities, more predictable physical work, lower levels of education, and relatively few social interactions are more susceptible to becoming redundant.

While we may find it hard to predict the talents and skills required in the future of work, we can be sure that professionals across all industries will need to nurture digital and human skills to effectively collaborate with rising digital technologies in the workplace. In addition, these elements will prepare people to access new occupations as they emerge or prepare themselves to transition to other jobs or careers, if necessary.

The Always-Employable **Worker**

AI and automation are on the cusp of revolutionizing the entire workforce. As they redesign existing jobs and create entirely new occupations, learning, adapting, reinventing, and building resilience to manage through instability are critical to succeed in an always-changing world. Focusing on these traits will increase your likelihood of employment, access higher-paid opportunities, and increase your job satisfaction. To successfully navigate the future of work defined by human-machine collaboration and become an "always-employable worker," you will need to:

1. **Master future-proof human skills** that are harder for machines to replicate; for example, developing the four Cs—critical thinking, creativity, communication, and collaboration. Over time, these skills will hold their value and help you grow personally and professionally.

2. **Become digitally fluent** to work in a digital and automated environment. Data analysis, digital collaboration, and digital ethics will become increasingly critical if you want to thrive in the digital world and adapt to the new ways of working.
3. **Develop cross-domain skills** by mastering more than one area or occupation to produce new outcomes and create value for yourself, your organization, and your customers.
4. **Learn how to manage higher** stress levels from constant changes in the future job market.
5. **Become a lifelong learner** by updating your skills and reinventing yourself continually.

Mastering Future-Proof Human Skills

By 2025 the amount of time humans and machines spend on tasks at work will be equal. As we share responsibilities with AI, we can retain our advantage in critical and strategic areas such as managing, advising, decision-making, negotiating, and engaging, while machines and algorithms focus on crunching data, recognizing patterns, and performing repetitive and manual tasks. The demand for human-centric skills will grow in the future of work. No matter the job title or industry, these skills can add value beyond what automation or intelligent machines can do. Honing human skills will make professionals more flexible, adaptable, and employable throughout their careers.

Some professionals will leave their existing jobs to pursue more meaningful work, shift careers, or become entrepreneurs. Likewise, automation or different business strategies will force other professionals to transition to new opportunities. Whether your career switch is intentional or unintentional, you will need to master future-proof skills to flourish in a changing environment.

Communication, critical thinking, and collaboration are abilities that have differentiated humans for thousands of years. As illustrated by Yuval Harari in his book, Sapiens: *A Brief History of Humankind*, the cognitive revolution has enabled Homo sapiens to think in unprecedented ways and communicate in a language that not only conveys information but creates imagined worlds. Thomas Edison left school at a young age. Still, he used the technology available at the time alongside skills such as creative thinking and persistence to advance humanity's progress, which allowed him to become one of the most important innovators of all time.

As we digitize essential aspects of our lives, including work, interpersonal communication will become even more critical as we navigate more ambiguous scenarios in the future. By enhancing your communication skills, you can effectively exchange information with other people and eliminate unnecessary misunderstandings, conflict, and frustration in your personal and professional lives. In addition, respecting and empathizing with other people and perspectives are critical abilities during the twenty-first century; becoming an active listener will be as important as being a dynamic speaker.

Human skills like adaptability, critical thinking, creativity, and emotional intelligence will remain vital to thrive in a volatile, uncertain, and ambiguous job market. These skills will not become obsolete or redundant; instead, they will serve as key differentiators for people in the digital economy. As jobs and tasks are divided between humans and machines, you will need to develop skills that are harder for a machine and robot to mimic. In the *Future of Jobs* report from the World Economic Forum, only two out of the top fifteen skills listed were directly related to technology expertise: "technology use and control" and "technology design and programming." The other thirteen skills were linked to critical and analytical thinking, active learning, problem-solving, creativity, stress tolerance, and flexibility.

AI, machine learning, and automation make the future of work a fast-approaching reality. As these systems transform every aspect of our lives and jobs, they will serve as vehicles to augment our human capabilities and help us become better predictors, evaluators, thinkers, and data analysts. Still, they will not replace human skills such as conflict resolution, social awareness, empathy, integrity, and compassion. Our human ability to empathize and collaborate with others will remain essential to building teams, relationships, and trust.

As professionals interact increasingly with machines and rely on their computing power and accuracy in many types of calculations, estimations, and predictions, they will need to become more familiar and comfortable working with AI systems. AI, robots, and chatbots can provide answers. Still, to collaborate effectively with machines, humans will need to ask the right questions and point them in the right direction, especially in areas like data analytics, ethics, and diversity. Becoming self-aware by actively listening, seeking relevant information, synthesizing data trends, and eliminating biases will help you ask the right questions, structure complex and unforeseen problems, and make effective decisions.

People can learn and transfer future-proof human skills from time to time or from one job or industry to other experiences. As the future of work unfolds, cultivating human skills will increase your employability. In addition, nurturing your "humanness" throughout your career will protect you from being replaced by automation or robots. By their nature, there is no manual, lecture, or desk procedure to learn unique human skills; you acquire them through constant practice and lifelong experiences. Therefore, it is vital that you proactively identify the critical abilities you need to improve in your skillset, routinely practice them, and remain flexible to adapt to unforeseen scenarios in the years ahead.

Develop Digital Fluency

Becoming digitally fluent can help you unleash creativity

and use digital technologies in empowering and transformative ways. As AI, machine learning, and social media networks embed themselves in all aspects of our lives, we need to develop the ability to use these tools and understand how they work. The ability to explain why they operate the way they do and navigate virtual spaces safely, collaboratively, and ethically will give you the confidence and skills to thrive in a digitally connected world.

According to the study "Skills Shift: Automation and the Future of the Workforce" prepared by McKinsey, automation will accelerate the shift in required workforce skills. The demand for digital skills will rise by 55 percent by 2030. Digital skills go beyond the basic capacity of searching the internet, sending emails, or attending virtual meetings. UNESCO broadly defines digital skills as a "range of abilities to use digital devices, communication applications, and networks to access and manage information. They enable people to create and share digital content, communicate and collaborate, and solve problems for effective and creative self-fulfillment in life, learning, work, and social activities at large."

As the nature of work evolves, working confidently with technology and developing a broader knowledge of AI, data analytics, and digital platforms can help you upskill and elevate your distinctiveness in the job market. Additionally, digital fluency allows you to develop and deploy digital systems, rather than merely using them or, at worst, being replaced by them. According to research conducted by the United Kingdom government, workers can reduce their risk of automation by 59 percent by having specific digital skills. "Specific digital skills commonly complement uniquely human skills such as design, writing or communication, which in combination are difficult to automate and critical to a firm's success," as stated in the report. "These specific digital skills are not solely required in the tech sector but are in demand across all sectors of the economy."

As the demand for digital skills continues to increase over the next few years, businesses and professionals must work urgently to reduce the digital gap and develop skills they will need to thrive in the future. Based on McKinsey's "Beyond hiring" survey, nearly 90 percent of executives are either experiencing skill gaps now or expect them within a few years. Professionals in technical and non-technical roles will need to converge their primary specialty with higher levels of digital fluency and become comfortable collaborating with technologies. Ensuring people develop high-demand skills and are prepared to use digital technologies, leverage collaboration tools, foster digital ethics, and develop critical thinking to make better decisions could result in significant economic growth. According to the World Economic Forum, closing the global skills gap could add $11.5 trillion to global GDP by 2028.

Cross-Domain Skills Growth

Developing cross-domain skills will help you navigate different disciplines and master a range of fields rather than being limited to a narrow or specific domain in the future job market. The digital world allows companies to explore new scenarios and create incredible opportunities at an accelerated pace. This trend will require entrepreneurs and professionals at all levels to nurture skills that go beyond specialization.

As digital technologies blur the lines between industries, acquiring cross-domain skills will allow you to explore more opportunities in different areas. As the nature of jobs evolves, interactions, processes, virtual spaces, and data will become increasingly complex and ambiguous. Professionals will need to weave a combination of specialized abilities with human and digital skills to reach new markets and audiences. For example, some lawyers are becoming experts in data privacy and digital ethics to attract and help more clients seeking support against cybercrime threats or data breaches. Doctors, nutritionists, and other subject matter experts are becoming digital content creators. By navigating social media networks and acquiring digital marketing skills, they are leveraging the massive potential of online platforms to reach larger audiences beyond their day-to-day activities.

As we navigate the digital era, in which we collaborate with sophisticated, robust, and more intelligent technologies, *knowing* will need to give way to *exploring* and *learning*. Narrow specialists, hesitant to understand and accept new technologies, processes, and business models or refusing to integrate concepts across disciplines, will struggle to remain relevant and competitive in the marketplace. In the years ahead, a cross-domain approach will be vital. Professionals will need to think through different disciplinary approaches. A cross-domain approach allows people to learn and translate knowledge and ideas to different contexts and develop abilities from other parts of the organization or economy. Remaining curious and open to continuous learning throughout your life will be crucial as navigating multiple jobs, disciplines, and even industries will become the norm in a digitally enabled economy.

In-depth knowledge in a specific domain will open doors to meaningful and high-quality jobs, but as jobs rapidly transform, developing skills in more than one domain will allow you to access new possibilities and experiences. As David Epstein states in his book *Range,* "The more constrained and repetitive a challenge, the more likely it will be automated, while great rewards will accrue to those who can take conceptual knowledge from one problem or domain and apply it in an entirely new one."

Managing Higher Levels of Stress

Stress management is one of the most critical skills in the twenty-first century. Unfortunately, American and Canadian workers ranked highest for daily stress levels, according to Gallup's "State of the Global Workplace" report. Roughly 57 percent of workers from these two countries reported feeling stressed daily, compared to 43 percent globally. Jim Harter, Gallup's chief workplace scientist, points out five elements that can help individuals thrive: career well-being, social well-being, financial well-being, physical well-being, and community.

Active professionals in the workplace and young people preparing to enter the job market will need to prepare themselves to cope and recover from ongoing changes. One way to reduce concerns about the future regarding AI, automation, and robots taking over human jobs is to learn and recycle skills continually. Professionals who make learning a habit will remain employable, regardless of how jobs are reconfigured or displaced. The need for ongoing learning is also noted in the "Future of Work" report prepared by Forrester: "As much as companies must become learning institutions, so must employees become learners—learning core skills, adapting to new working models, and understanding what it means to be ready and fit for the future."

Improving employees' well-being positively impact their personal lives and productivity and benefits companies' overall financial performance. For example, Harvard researchers concluded that, on average, for every dollar companies spent on wellness programs, their healthcare costs fell by approximately three dollars. In addition to self-improvement and frequent recycling your skills, you need to revisit and maintain healthy habits and behaviors. For example, wellness and healthy eating and living are your best weapons against workplace stress in the fluid future of work.

Become a Lifelong Learner

To operate in a highly dynamic working environment, you will need to continuously learn and develop skills across domains, enhance your future-proof human skills, and become digitally fluent. As existing jobs and careers change to something new, self-development and lifelong learning will be prerequisites to access meaningful work in the future. Lifelong learners educate themselves throughout their lifetime. They focus on fostering continual learning and education to develop skills needed for high-quality employment and personal fulfillment.

Professionals will need to move away from their comfort zone because stagnation or a lack of challenges to sharpen their abilities and expertise will likely make them redundant. Individuals who move away from their comfort zone and embrace an exploration zone—a state in which people are comfortable with unfamiliar scenarios and curious to explore new skills, activities, and connections—can create value for themselves, their organizations, and the community. Learning is not a separate workstream or an extra effort for lifelong learners. Instead, it happens naturally. They take on challenges, learn from mistakes, and actively seek new knowledge and abilities. Although they experience rough days like anyone else, lifelong learners glean more from daily opportunities because everything—experiences, relationships, meetings—presents an opportunity to develop and grow personally and professionally.

Education Doesn't End **with a Diploma**

Our current education system, built on the Industrial Revolution model, fails to prepare people to navigate an always-changing workplace. The one-size-fits-all approach needs to be replaced by flexible, on-demand, and individually customized learning environments. Instead of memorization and standardization, people need to be empowered to solve complex, real-world, and global scale problems. Thanks to the exponential growth of AI, mobile phones, and online learning platforms that offer a host of solutions and virtual experiences, we can learn and develop in-demand skills anytime and anywhere.

Peter Diamandis and Steven Kotler have introduced the six Ds of exponentials to contextualize technological change. According to Diamandis and Kotler, the growth cycle of digital technologies takes place in six key steps: digitization, deception, disruption, demonetization, dematerialization, and democratization. "The six Ds are a chain reaction of technological progression, a roadmap of rapid development that always leads to enormous upheaval and opportunity."

Innovations from edtech have the potential to digitize, demonetize, and democratize learning experiences for everyone. The widespread use of virtual and self-learning will increasingly provide significant value and on-demand learning opportunities for students and professionals to develop new skills and succeed in a competitive, dynamic, and digital workplace.

Learning Anywhere **at Any Time**

The high cost for secondary education in the United States is a key factor for a staggering $1.7 trillion student loan debt, the second-highest consumer debt category, following mortgage. Although college graduates earn weekly wages up to 80 percent higher than those of high school graduates, the overall dropout rate for college students is about 40 percent, with approximately 30 percent dropping out before their sophomore year. These numbers suggest we need to investigate why so many students drop out and what we can do to enable more people to obtain degrees and successfully prepare for a rapidly changing workplace.

AI, virtual reality, cloud computing, and online learning will radically transform education design and delivery. These technologies will make learning more interactive, immersive, and engaging. The benefits of e-learning are vast. It can help increase engagement levels, reduce education costs, and provide more engaging ways to develop high-demand skills. E-learning is poised to transform and improve learning experiences radically. For example, it takes 40 to 60 percent less time to complete when compared to traditional learning, particularly as students can learn at their own pace, accelerate through concepts as they wish, and select the most convenient time to study.

As the world changes and technology continues developing, it becomes increasingly important to stay informed, trained, and aware of the key trends shaping jobs and industries. Companies will want their workforce to refresh their skills and abilities faster and more frequently. Seventy-three percent of employers agree they will lose ground to competitors if they don't reskill their employees. Massive open online classes (MOOCs) will become more broadly available. They will serve as important tools for professionals to quickly educate themselves, develop skills to thrive in the future, and help companies close the skills gap.

Companies, learning platforms, and schools can use virtual reality and augmented reality for highly experiential and immersive forms of learning. They offer a higher level of student engagement compared to traditional methods like learning based on listening or reading. According to the United States National Training Laboratory, the retention rates for training through virtual reality are 75 percent, far above the 10 percent for reading-based learning and 5 percent for lecture-style education. These technologies can create virtual environments and expose students to a wide range of realistic experiences by visualizing concepts and learning new skills interactively. As virtual reality gains traction in the years ahead, traditional physical spaces like classrooms or meeting rooms will transition to always-on virtual learning settings.

AI, machine learning, virtual reality, and online learning platforms will allow education to become highly personalized as they will increasingly understand, collect, and analyze students' data and performance. As these technologies identify students' strengths and weaknesses, they can tailor the material to each individual's preferences and needs and make the learning process more enjoyable and effortless. The future of learning will be more flexible, on-demand, and customizable; this will enable lifelong learners to access more meaningful education to develop high-demand skills.

Getting Ready for **The Future of Work**

While digital technologies will reconfigure some of the existing jobs and eliminate others, they will also create novel opportunities in the future. The individuals actively tracking the technologies shaping industries and careers and recycling their skills will position themselves to access high-demand and well-paid jobs. The secret to succeeding in the jobs of tomorrow is to be intentional about learning and, more importantly, not to fear the rapid advances of AI and automation. Our biggest worry should be inaction and stagnation. Thanks to the internet, AI, and online learning platforms, we can master human and digital skills to become an "always-employable worker" in the future.

We need to become better day in and day out and continually develop new skills to adapt to any unforeseen situation, including transitioning to other jobs or even careers, if needed. In an article published by the World Economic Forum, Karin Kimbrough, chief economist for LinkedIn, states: "Our research reveals the majority of transitions into jobs of tomorrow come from non-emerging jobs, proving that many of these jobs are more accessible than workers might think." Consistent learning and willingness to embrace change are essential to enable professionals to switch from declining jobs to areas in increasing demand. According to the World Economic Forum *Future of Jobs* report, 50 percent of career transitions into data and AI are from non-emerging roles. That figure is much higher for sales roles (75 percent), content creation and production positions (72 percent), and engineering roles (67 percent). This trend will continue as we move closer to 2050. To prepare yourself, keep learning!

● ● ● ● ● ● ● ● ● ● ● ● ● ● ●

A 2050 Scenario: **A Dynamic Job Market**

Noah has transitioned careers and industries multiple times throughout the last three decades. Today he is a robotic rapport coordinator for a manufacturing company specializing in producing flying vehicles. Flying cars seemed immensely futuristic to Noah thirty years ago, but now it's an industry worth over $1 trillion and one of the most efficient transportation modes. Significant advances in AI, satellite communications, ultra-efficient batteries, and urban infrastructure allow autonomous cars to fly people and goods. In addition, the flying vehicles use rotors instead of wings, which allows for vertical takeoff and landing. Flying cars are transforming the way people commute, work, and live.

In 2021, Noah's job didn't even exist, and most of the flying car projects were in the conceptual phase; what sounded like sci-fi back then is now dominating the skies in most larger cities around the world. As a rapport robot coordinator, Noah supervises humanoid robots in the production line. Besides driving productivity increases, he needs to ensure the robots have harmonious relationships with human workers. Job descriptions have changed so much in the last decades that to survive such a dynamic and disruptive job market, Noah needs to focus on mastering and to combine a range of technological, emotional, human, and digital skills. Developing these abilities helped him add value beyond what can be done by robots, automated systems, and intelligent machines. Leveraging his future-ready mindset, intuition, and creativity allowed him to continually adapt to new ways of working and create solutions to improve people's lives worldwide.

Although he is nearly thirty years older than his daughter, Eva, both leverage technologies to enhance human abilities and experiences. While Noah trains humans and robots to work collaboratively, Eva uses digital technologies to bring people and society together. When Eva was a little girl, her dream job was to save Planet Earth; now, in 2050, it has become a reality. Sustainable world regulator is her new job title. The waves of change are making problem-solving and decision-making increasingly complex. Her broader understanding of the world alongside ethical principles, empathy, and collaboration spirit allowed Eva to effectively make sense of the rapid technological, environmental, financial, and societal changes around her—topics the traditional educational system failed to teach people in the past centuries.

Leveraging billions of new users joining the internet in recent years, she is convinced the digital world can connect the people. Through her work in the Go Human movement, she has helped millions of young individuals worldwide rehabilitate and rebuild their lives in the physical world in the last decade. As a climate change steward, she has helped citizens become more conscious of global warming and how their actions can solve the net-zero equation. She has also helped many countries to end and reverse deforestation. Through her fundraising and collaboration with the private and public sectors and citizens, her project has helped plant, protect, and restore trillions of trees worldwide. Eva is a sports fan, and in her free time, she's a coach's assistant for the national humanoid robot basketball team. Her main priority is to enable robot athletes to enhance their physical traits, similar to human athletes. She is preparing her team to compete in the 2050 Olympic Robot Games.

Work as we knew it thirty years ago has changed. Fixed

employment contracts, working for one company, and defined job structures and responsibilities have gone away. AI, machine learning, robots, IoT, and virtual reality made traditional working environments obsolete. Technology has created new jobs, but automation has caused structural unemployment in many parts of the world. Many occupations in the transportation, financial, manufacturing, and professional services sectors are executed by machines, algorithms, and robots. The rapid technological disruption is redefining social norms and rules that we have adhered to for centuries. Some laid-off workers haven't been able to access alternative employment due to their inability to keep up with the accelerated changes in the marketplace and unwillingness to develop new skills; unfortunately, they still rely on universal basic income from governments. However, many people no longer chase paychecks. Instead, they collect experiences and work for a purpose. This shift has made people a lot richer and freed up their time for more fulfilling activities. Digital technologies and the internet have allowed people to invent their own jobs and bring their whole self to work instead of bringing a different version or personality to fit available positions. Skills are being developed at the same speed as new occupations are emerging, so building highly specialized expertise is giving way to adaptive learning, where people possess a range of diversified skills, including digital fluency and cross-domain abilities. Unique human skills such as critical thinking, creativity, communication, storytelling, integrity, and collaboration have become increasingly more valuable as people are forced to adapt to new ways of working and an always-changing digital economy.

If used **wisely, ethically, and broadly available,** digital technologies allow people to access quality **education, employment, real-time information, increase social engagement, and promote small businesses.**

9 The Role of Technology in **Addressing Inequality**

As the world moves rapidly toward new scenarios filled with unknowns, we need to address critical systemic issues around wealth distribution, gender disparities, access to digital technology, and affordable education so the world can enjoy a fairer and more equitable future.

Since the Industrial Revolution, technology has been a critical driver of economic growth and social progress. New technologies have made products and services more efficient and accessible, contributing to increased productivity and per capita income. Higher incomes and more consumer choices have contributed to a higher quality of life in many parts of the world. As emerging technologies such as AI, satellite internet, IoT, social media networks, virtual reality, online learning platforms, and many other digital technologies grow exponentially, they will increasingly offer benefits to reduce inequalities. As a result, people will have more opportunities to access higher-quality information, education, healthcare, financial services, employment, and increase their social engagement. However, if digital technologies aren't deployed responsibly, equally, and ethically, they also bring social and economic challenges and risk exacerbating inequalities.

AI, machine learning, and robots can make many workers redundant by automating rule-based, repetitive, and predictive processes. The disruptive and transformative potential of these technologies requires challenging transitions. Inevitably, it creates winners and losers. "A lot of the new job opportunities that technology brought from the 1960s to the 1980s benefitted low-skill workers," says MIT economist Daron Acemoglu. "But from the 1980s, and especially in the 1990s and 2000s, there's a double whammy for low-skill workers: They're hurt by displacement, and the new tasks that are coming, are coming slower and benefitting high-skill workers."

As AI, digital platforms, and automation grow exponentially in the years ahead; they will impact workers disproportionately. These technologies favor non-routine cognitive tasks while reducing the demand for manual and repetitive work. As a result, highly skilled workers, for example, executives, engineers, scientists, designers, and doctors, are better positioned to benefit from digital technologies' rapid expansion. Conversely, AI, robots, and chatbots can replace middle and lower working classes performing more predictive and manual activities such as machine operators, receptionists, administrative assistants, customer service representatives, and data entry clerks.

According to research published by the National Bureau of Economic Research, wage and earning inequalities have risen sharply in the United States and other industrialized economies over the last forty years. Workers with a postgraduate degree have seen their wages increase, while less-educated workers have seen their wages decline. The research claims that automation has been the primary driver of income inequality: "Between 50 percent and 70 percent of changes in the US wage structure over the last four decades are accounted for by the relative wage declines of worker groups specialized in routine tasks in industries experiencing rapid automation." Therefore, lifelong learning and adaptability are imperative to remain relevant in the workforce and reduce wage disparities in a fast-paced and digitally working environment. If technological progress is faster than progress in education, the growing demand for highly skilled workers may increase inequalities.

Technology can help reverse underlying socioeconomic issues aggravating income inequalities worldwide by providing people reliable access to information, equitable job opportunities, and cheaper online learning to develop in-demand skills. However, if significant changes are not made now, we could see even more significant disparities between the rich and the poor in the years and decades ahead.

The wealth gap between ethnic groups in the United States also demonstrates critical disparities. According to a Federal Reserve report, the typical white family has eight times the wealth of the typical Black family and five times the wealth of the typical Latino family. While white Americans hold nearly 85 percent of the nation's wealth, Black households represent just 4 percent. According to a McKinsey report, black workers tend to take supportive roles in established rather than high-growth industries in the job market. They are underrepresented in seven of the eight highest-paying industries and self-directed and creative roles such as software developers or managers. As a result, they are at greater risk of displacement by advancing digital technologies. Inequality is also reflected in the disproportionate employment rates between men and women. Women's global labor force participation rate is close to 49 percent, while men have a 75 percent participation rate.

Reaching a balanced scenario where one group is not favored over another will require significant actions and commitments from key stakeholders. Digital technologies and social media platforms can drive inclusion and equality by empowering people to become more aware, educated, and skilled to enjoy a purposeful life and career.

Using Technology's Help to **Build an Equal Scenario**

Although digital technologies can reduce the physical distance between individuals by flattening or removing boundaries between countries and communities, it is critical to ensure their potential is used as a social equalizer. Building an equal scenario is a daunting task as inequality is one of the most pressing issues of our time. There are significant ways digital technologies can help reduce disparities in the twenty-first century. However, the challenge of achieving a more inclusive society is highly complex, and technology cannot solve inequalities alone; instead, the responses need to come from solid collaborations between governments, private sector, and citizens. Nevertheless, technology can complement and support organizations and people's actions and decisions towards building a more equal, diverse, and inclusive world.

To avoid a scenario where a few fortunate winners dominate and manipulate a minority, we must progress toward equality and fair access to the internet, lifelong learning, and equal job opportunities regardless of socioeconomic background, gender, race, sexual preference, or culture. To ensure technology can help narrow the opportunity gap and create a more inclusive society, we need to address three core actions. Executing these initiatives will provide everyone a fair chance to fulfill their potential in the future:

1. Approach technology as a social equalizer tool.
2. Ensure equal access to high-quality jobs in technology.
3. Increase key stakeholders' commitment and engagement.

Using Technology as a Social Equalizer Tool

Digital technologies are increasingly transforming people's lives in many parts of the world. The social impact of AI, digital technologies, and social media platforms can be vast and positive. If used wisely, ethically, and broadly available, they allow people to access quality and affordable education and training programs, work on more fulfilling jobs, consume compelling content and information, and create an ecosystem to promote small businesses.

Today's children will be in their thirties by 2050; they must learn beyond literacy and numeracy; they need to seek human, transferable, digital, and entrepreneurial skills to compete and thrive in the future. As early childhood education and digital fluency can yield a significant return on investment later in life, the earlier we take consistent actions to increase their access to the internet, the higher our chances are to reduce inequalities and raise lifelong learners. Yet, according to a report from UNICEF and the International Telecommunication Union, two-thirds of the world's school-age children—or 1.3 billion children aged 3 to 17 years old—do not have internet connection in their homes. Among school-age children from wealthy households, 58 percent have an internet connection at home, compared with only 16 percent from the poorest families. As the world becomes more digitally dependent, ensuring children have equal access to the internet is crucial to reducing inequality. Otherwise, we run the risk of isolating them from the benefits coming from the digital economy in the future.

Technology companies need to prioritize eliminating bias from AI and machine learning training models used for sensitive areas such as hiring, loan approvals, medical diagnosis, and criminal justice sentences to avoid inappropriate decision-making affecting people's lives. Thus, while technology can make us healthier, more prosperous, and aware, AI systems must follow ethical principles to avoid unjust impacts on people's lives and well-being. Increasing everyone's general awareness and nurturing digital skills can help create a safer, fairer, and more responsible digital universe. People need to understand how emerging technologies can impact other individuals and society so they can securely engage, collaborate, and enjoy the benefits of the virtual world. Joy Buolamwini, computer scientist and digital activist at the MIT Media Lab, says: "The public is largely unaware of the ways in which AI shapes their lives, and there are few regulations that require disclosure about the use of the technology. Without awareness about the uses, risks, and limitations of AI, we remain at the mercy of entities that benefit from opaque AI systems, even when they propagate structural inequalities and violate civil rights and liberties."

As we aim to build a more transparent and equitable digital world, how can people increase awareness and prevent negative impacts from AI systems? One answer is to allow them to take complete control of their private data by letting them choose what personal and sensitive information should be retained or removed from a company's databases; for example, Dell Technologies is committed to providing more clarity and transparency around personal data; by 2030, the company will fully automate its data control processes, making it easy for customers to control their personal data. In addition, introducing critical topics such as AI biases and technology ethics to students' formal education teaches them the importance of technology principles, data privacy, and human agency. By learning these topics earlier, young people can equip themselves to navigate a digitally-enabled world. Additionally, increasing women's participation in STEM occupations allows them to access higher-quality jobs, narrow the gender pay gap, enhance women's economic security, and bring diverse perspectives to research and the development of digital technologies.

The benefits of the digital world must outweigh the negatives. Digital technologies, AI, and social media platforms will advance rapidly; it's important to limit how these companies can influence our daily lives, culture, and economy. With the right level of accountability, policies, and regulations, we can navigate a safer, less biased, and harmful digital world while accelerating economic growth and addressing rising inequalities.

Enhancing Equal Job Access in Technology

We face a long, uphill journey to achieve gender equality in the workplace. According to the United Nations, women held only 25 percent of seats in national legislatures in 2020; although that's twice the level in 2000, the gap is still significant. In addition, women held one-quarter of corporate managerial positions; and accounted for less than one-tenth of the CEOs of major companies. To build an equal and diverse workforce, companies need to hire professionals who can add value, have a positive attitude, and are willing to reinvent themselves continually. Therefore, no restriction should be made beyond proficiency and suitability to select a candidate. Race, gender, sexual identity, ethnicity, or physical disability should not prevent people from accessing work opportunities or advancing their careers and education.

No matter their background, people need to be equally equipped with digital and human skills to thrive in an ever-changing business landscape. One way to mitigate gender inequalities is ensuring women receive education and hold the same number of jobs as men in the labor market; this could potentially boost the global GDP by $20 trillion by 2050, based on a study from Bloomberg Economics. According to an analysis, "Policy changes that result in more women entering the workforce, like those that bolster female access to secondary education, childcare, and flexible work arrangements, have the potential to "light a fire" under global growth over the next three decades." Ensuring women have access to quality and affordable learning experiences to develop diverse and high-demand skills will also contribute to equal opportunities in the future workplace.

Women are underrepresented in jobs that require advanced technical skills. In data and AI, women make up 32 percent of the workforce; in engineering, 20 percent; and in cloud computing, only 14 percent, according to the World Economic Forum's Global Gender Gap Report 2021. Women must have access to ongoing upskilling and reskilling opportunities to equip themselves with essential AI, cloud computing, cybersecurity, and data science skills to fulfill high-quality jobs and leadership positions instead of being hired exclusively for lower-level and execution roles. They need to be engaged in strategic planning, decision-making, and creative projects, including developing cutting-edge technologies that will reinvent industries and reshape the world in the coming years and decades. Equipping women with digital and human skills will lead to more diverse teams, which helps catalyze innovation, business growth, and more ethical and responsible systems. According to a survey from Deloitte, 71 percent of the respondents strongly agreed that adding women to AI and machine learning will bring unique perspectives to technology. "A more diverse workforce is better equipped to identify and remove AI biases as they interpret data, test solutions, and make decisions. Specific to gender, women are likely to catch things men might miss (and vice versa). In this regard, gender diversity can benefit AI development."

To meet the technology industry demand for qualified professionals, women need to acquire the right in-demand skills to pursue the jobs yet to be created. Therefore, the public and private sectors must work together to increase women's access to computer science and high-quality programs that connect them to STEM education and careers.

Key Stakeholders' Commitment and Engagement

Technology can help mitigate inequalities and serve as an essential tool to empower individuals and enable the public and private sectors to solve problematic global issues around income gaps, racial injustice, healthcare, climate change, hunger, and food insecurity. As new technologies evolve exponentially and fast adoption drives business growth, solutions for an equal workplace and society will need to come from an integrated framework of three main pillars:

1. **The public sector** must be the catalyst to enhance equality by introducing and maintaining focused policies and strategies to ensure new technologies are readily accessible and helpful for everyone at any time. We cannot risk leaving behind a large portion of the global population simply because they do not have access to an affordable and reliable internet connection. AI, online learning, and virtual reality hold the promise of a more prosperous future for all, but effective regulation needs to be in place for us to realize that promise. Policies need to regulate the rapid proliferation of new digital technologies and remain responsive to change as they radically reinvent jobs and industries. Notably, the profound impact of AI systems and automation needs to account for people's well-being, reskilling, upskilling, and long-term prospects. Hence, the public sector needs to ensure people receive training programs to develop high-demand skills or relocation support to find new jobs in case of displacement by automation. If some workers shift to freelancing and the gig economy, providing benefits and protections like full-time employees can create a financial safety net for this burgeoning workforce. Governments need to develop better frameworks to allow everyone a chance to succeed, including investing in development and training programs to create a gender-equitable workforce. Creating new career pathways for women and advancing equity in technology research and development can boost the individual quality of life and the economy as a whole. For example, one study found that innovation in the United States would quadruple if women, people of color, and children from low-income families innovated at the same rate as other groups who are not held back by discrimination and structural barriers.

2. **The private sector** is a critical engine to promote equality in the workplace. Companies need to spur inclusiveness, transparency, and trust to ensure that everyone gets equal opportunities regardless of race, gender, sexual orientation, citizenship, or disability. Businesses must foster diverse people, perspectives, and opinions and make bold decisions to build an environment where everyone feels included. From now until 2050, diverse perspectives, ideas, and teams will be key competitive differentiators in the marketplace. The private sector must provide its workforce with continual learning experiences and future-proof skills, particularly ensuring employees have opportunities to improve digital fluency and use digital technologies in empowering and transformative ways to thrive in the future. Employers need to expose their people to new experiences, abilities, and opportunities so they can feel more comfortable with change; this makes employees better equipped to adapt to a constantly changing workforce and increases the pool of skilled, diverse, and versatile talents for companies to use in different projects. With the vast potential for AI to transform society and the global economy in the coming years, companies designing these systems must eliminate biased algorithms and ethical concerns from their technology. For example, leading technology companies like Microsoft, Google, and IBM have defined specific principles to increase transparency and ethics around their AIs' design and deployment. According to IBM, the purpose of AI is to augment human intelligence, "AI should make all of us better at our jobs, and that the benefits of the AI era should touch the many, not just the elite few." Microsoft shares similar principles, stating, "Everyone should benefit from intelligent technology, meaning it must incorporate and address a broad range of human needs and experiences. [. . .] AI systems should treat everyone fairly and avoid affecting similar situated groups of people in different ways." Google's objectives for AI applications are to "be socially beneficial," "avoid creating or reinforcing unfair bias," and "be accountable to people."

3. **Most importantly, people.** We are responsible for progressing our lives and shaping our future. Personal initiative, commitment, and focus on long-term benefits are crucial for any individual who wants to thrive in a world defined by discontinuity. Self-starting people proactively prepare themselves emotionally, physically, and financially to overcome barriers and setbacks. They improve their skills and qualifications to access more fulfilling and rewarding opportunities in the future. As the future of work is built on a foundation of lifelong learning, we need to be aware that AI, automation, globalization, new market entrants, and shifting demographics will drive job instability. Therefore, we will need to update our abilities, perspectives, and assumptions more frequently to respond to the rapidly-changing skills requirements and business models transformations. Considering this, learning at every stage of our life is imperative to access the fastest-growing and higher-value jobs in the future. As the world of work evolves and jobs become more digitally focused, digital skills will become increasingly crucial for jobs across all industries. Preparing to access the jobs of tomorrow by cultivating digital and human skills will avoid job displacement.

A 2050 Scenario: **Women Helping Women**

Substantial progress has been made to mitigate gender inequalities, particularly by ensuring women have equal access to education and job opportunities as men. However, some countries have advanced more rapidly than others towards gender parity.

Eva has grown up inspired by her great-grandmother Maria's life story. Maria, an immigrant from South America, faced unemployment and financial hardship. Her parents decided to seek better living conditions in North America. These adversities never prevented Maria from finding ways to pursue a better life and prosperity. After World War II, the world was reconstructing itself, and jobs were not as easy to find as her family had hoped. As an immigrant, she fought to conquer her space on the job market. Working harder than some of her more privileged peers alongside personal initiative, commitment, and ongoing education allowed her access to meaningful job opportunities.

Eva is applying the values she has learned from her great-grandmother. By understanding the importance of empathy, sociability, and community, she is committed to ensuring AI is used to unite people instead of increasing the gap between the richest and poorest. As a forward-thinking and global citizen who believes digital technology is a social equalizer tool rather than a divider, she wants to reduce gender inequalities in the workplace. She has been working with young women in marginalized communities worldwide to prepare them for the fastest-growing jobs in the twenty-first century. Through her Go Human movement, she exposes girls to advanced technologies and projects to develop in-demand skills at the same pace as young women in more developed communities.

Recently, Eva launched a new project under the Go Human movement, a global competition to accelerate innovation by using the power of AI and machine learning to solve real-world problems. Out of thousands of participants, the winner was a young woman who built an AI-powered digital assistant that helps single parents care for their children. The digital assistant alleviates the burden of home and family care by allowing parents to fully automate household activities, from scheduling appointments, managing family calendars, buying groceries online to helping children complete school assignments and stay physically active. As a result, the technology reduces the stresses of balancing home and work obligations while freeing up parents' time to develop skills to remain relevant in the workplace.

Eva's efforts are creating a positive impact in the life of many girls. But she is not alone. Companies in many parts of the world have championed diversity, equality, and inclusion in the last decades. By committing to creating learning environments where young women are encouraged to develop digital skills at the same level as young men, these companies were able to achieve a 1:1 ratio of women to men in critical technology jobs like researching, designing, and programming technological systems. Closing the gender gap in the workforce is helping to improve women's motivation and morale and bring trillions of dollars to the companies and countries that promote early awareness and education to mitigate gender disparities. Countries that actively worked to close the gender gap by ensuring women and men were treated with equity saw the quality of life of their people and social engagement radically improve. However, they still need to work very hard to address the gap in intersectionality.

Today, we are witnessing the beginning of a **new era of space exploration driven by humanity's curiosity** to explore the unknown, discover new possibilities, and push the boundaries of our **scientific and technological limits.**

10 Exploring Beyond the Earth

We are living in a time where much of what used to be considered science fiction is becoming a reality, primarily due to the significant advances in scientific exploration and technologies that are transforming every aspect of our lives. The new age of unprecedented innovations is challenging our assumptions and perspectives; as emerging technologies and science evolve rapidly and disruptively, they will redesign how we live on Planet Earth and accelerate humanity beyond our terrestrial home.

As scientific and space technology developments advance in the coming years, Earth will be borderless. Space exploration will be an increasingly important leap forward for humankind to address global challenges. For decades space exploration has transformed and enhanced the quality of life on Earth. Satellites in space have expanded internet access to billions of people, including bringing services to rural and remote locations. Insights generated from the space have enabled highly accurate GPS navigation and accurate weather forecasting. Experiments on astronauts in space have improved our understanding of medical conditions on Earth, including findings to treat people suffering from cardiovascular disorders, diabetes, osteoporosis, and balance problems. In addition, space exploration inspires younger generations by sparking their interest in STEM fields.

The race to space in the twenty-first century is already happening. Space programs and cutting-edge technologies previously—and strictly—managed by governments and highly specialized agencies such as the National Aeronautics and Space Administration (NASA) are now expanding to new horizons. Blue Origin's vision is to enable a "future where millions of people are living and working in space for the benefit of Earth." In addition, Jeff Bezos, founder of Blue Origin, has an environmental vision for space exploration: "We need to take all heavy industry, all polluting industry and move it into space, and keep Earth as this beautiful gem of a planet that it is." SpaceX has already carried astronauts to the International Space Station and launched and returned two NASA astronauts with its spacecraft for the first time. This achievement made SpaceX the first private company to send people into orbit.

Space exploration will continue to expand in the next decades, allowing more countries, companies, and people to explore new possibilities beyond Planet Earth. According to Benji Reed, director of crew mission management at SpaceX: "We want to send all kinds of people to space. [. . .] Everything we're doing is to open that new chapter in the space age." The investments and advanced research underway have the potential to continue lowering the cost to access space. Commercial launches have reduced the cost of sending a kilogram of supplies to space by a factor of twenty; NASA's space shuttle had about $54,500 per kilogram, but for a SpaceX Falcon 9, the cost is just $2,720. More affordable launches will accelerate space exploration and settlement. The increasing interest and efforts to expand space exploration will benefit humanity beyond offering a short flight or allowing people to cruise in a luxury liner that can float above Earth's atmosphere. The benefits of space exploration can be categorized as direct: the generation of scientific knowledge, diffusion of innovation and creation of markets, and inspire people around the world; or indirect: tangible enhancements to the quality of life such as improved economic prosperity, health, environmental quality, safety, and security.

According to Morgan Stanley, the global space industry could generate more than $1 trillion by 2040, and half of the projected growth of the space economy is estimated to be driven by satellite broadband. Audrey Choi, Morgan Stanley's Chief Sustainability Officer, has said, "satellite technology and space exploration offer a potential new frontier of opportunities to assess and address climate change and sustainability on a global scale." Additionally, the growing investments in satellite infrastructure are critical to expanding faster and more stable internet access to the farthest corners of Earth. Expanding network technologies like 5G is vital to unlocking the full potential and data flows from other exponentially growing technologies such as self-driving cars, robots, and big data. From now until 2050, data captured in space will increase in value, volume, and variety.

Cooperation between countries and governments to speed up space exploration is increasing. More than one hundred countries have signed the United Nations' Outer Space Treaty, which states that international law applies: "Outer space, including the moon and other celestial bodies, shall be free for exploration and use by all States without discrimination of any kind, on a basis of equality and in accordance with international law, and there shall be free access to all areas of celestial bodies." The treaty goes on to require that the "exploration and use of outer space, including the moon and other celestial bodies, shall be carried out for the benefit and in the interests of all countries, irrespective of their degree of economic or scientific development, and shall be the province of all mankind."

Besides advancing science and technology to help us solve critical global challenges and inspire younger generations to imagine a future with limitless possibilities, space exploration is also fostering diversity and inclusion. As part of the development of the modern human landing system that will carry the next two American astronauts to the moon's surface and pave the way for lunar exploration under the Artemis program, at least one of those astronauts will make history as the first woman on the moon. "NASA and our partners will complete the first crewed demonstration mission to the surface of the Moon in the 21st century as the agency takes a step forward for women's equality and long-term deep space exploration," states Kathy Lueders, NASA's associate administrator for Human Explorations and Operations Mission Directorate. "This critical step puts humanity on a path to sustainable lunar exploration and keeps our eyes on missions farther into the solar system, including Mars."

As space exploration and technology continue to expand in the next decades, so too will our ability to reach new horizons and deeply explore all aspects of the solar system, particularly the moon and Mars. NASA is working with American companies and international partners to uncover new scientific discoveries and lay the foundation for private companies to build a lunar economy. The agency wants to use the information and learning from exploring the moon to prepare humanity for the next giant leap—sending astronauts to Mars. The moon provides an opportunity to test new tools, instruments, and resources that could be used on Mars, including human habitats, technologies, and practices that could help us build self-sustaining outposts away from Earth.

The Multiplanetary **Human Reset**

Thousands of years ago, Homo sapiens had to gather and hunt food. As a result, they evolved behaviors and abilities to help them respond to survival challenges in unstable environments, inadvertently creating a turning point in the history of life on Earth.

In addition to hunting animals, they began to use Earth's resources in new ways to construct settlements. Is the destiny of humankind bound to Earth? As humans advance space exploration, will we see a new era of mining or manufacturing on another planet?

While it's still unclear if we will inhabit Mars in the future, the red planet continues to captivate humanity's curiosity. A trip to Mars would take anywhere between six and nine months based on today's technology; however, Elon Musk, founder of SpaceX, remains focused on establishing a permanent human presence there: "It's been almost half a century since humans were last on the moon. That's too long, we need to get back there and have a permanent base on the moon—again, like a big permanently occupied base on the moon. And then build a city on Mars to become a spacefaring civilization, a multi-planet species."

Over recent decades, humans have been trying to discover what Mars can teach us about how planets grow and evolve and whether it has ever hosted alien life. NASA's Perseverance rover is exploring the surface of Mars, assessing conditions for a possible human landing. One of their key missions is to find subsurface ice that could be used as a water source. Several crewed missions are being planned for the next decade; NASA's target is to land the first humans on Mars sometime in the 2030s. To continue boosting its space ambitions in a battle with the United States, China aims to send its first crewed mission to Mars in 2033 to build a base. The United Arab Emirates became the first Middle Eastern country to reach Mars; their one-hundred-year plan aims to establish the first human colony on Mars by 2117.

Over the last few centuries, new infrastructure and technologies have significantly reshaped the world, thanks to the relentless human beings whose mission was to make Planet Earth a better and more connected place. Factories, electricity, airplanes, railroads, and the creation of the internet have been instrumental in driving humanity's progress. Today, we are witnessing the beginning of a new era of space exploration driven by humanity's curiosity to explore the unknown, discover new possibilities, and push the boundaries of our scientific and technological limits. The advances of space exploration, particularly addressing whether life ever existed on Mars and the search for resources like water to create an Earthlike environment for human habitation on Mars continue to test the limits of our imagination. However, once critical technological, scientific, and engineering challenges are solved, we will know whether there's a future for our species and a solid commercial space economy on another planet.

A 2050 Scenario: **Above and beyond Earth**

Noah, Olivia, and Eva have decided to visit Mars. It's their first trip to space; they want to see and experience humans' progress on the red planet over recent decades.

As they travel to Mars, they are astonished by the breathtaking views of Earth from space. During the past thirty years, significant progress has been made to allow humans to land safely on Mars. However, people must wear highly specialized equipment to protect themselves from dust storms and radiation when they step onto the surface. This equipment also minimizes the impact of atmospheric pressure and temperature swings. Inside the resort, though, they quickly forget that they are nearly thirty-five million miles away from their home planet. The resort's facilities provide a similar experience to a beach hotel in Cancún, including a variety of swimming pools, spas, tennis courts, and restaurants.

The resort has plenty of infrastructure to produce enough oxygen, water, and energy for all guests. It replicates an Earthlike ecosystem so humans can stay there safely. The resort's facilities have green spaces and outside areas where guests can enjoy the sunlight without being affected by high radiation levels. Crops grown with hydroponics feed the visitors at the resort. 3D printers produce most materials and resources since round trips to Mars are still expensive. All facilities have technology that controls the temperature, cleans the air, and removes potentially harmful substances. At night resort guests enjoy shows and a casino where gamblers use Martian cryptocurrency.

Besides enjoying this exotic travel destination, Noah, Olivia, and Eva were intrigued to learn more about the social structures defining Mars's colony. Many elements governing Planet Earth, such as politics, laws, and nationalities, do not yet exist on the red planet. As humans settle on Mars, many social, economic, and trade policies are being rewritten to fit the unique environment. More important than developing a set of standards to protect Martian workers, the local laws are being created to ensure that resources like water, breathable air, and protection from radiation are available to everyone. A new political system is established to avoid inequality and unfairness; people have a voice and rights instead of a radical leader or group holding all the power. Seizing power in such a vulnerable place would be simple—control the available oxygen, retain water containers, or block citizens from returning to Earth. Ensuring there is a common ground for humans to relate to one another and leveraging local resources for survival are the most critical priorities for humans as they colonize Mars.

As humans settle on this new planet, a new working economy emerges to support the growing population. The economy is self-sufficient, using locally sourced materials and recycling as much as possible due to the distance and cost of receiving goods from Earth. Although the travel time has significantly reduced, it still takes months to transport supplies from Earth to Mars. The first jobs created were related to agriculture—growing plants for food—and finding efficient methods to distribute water. As colonies and cities grow, real estate is becoming a hot market. Families are expanding, more residents are moving from Earth, and the need for new homes increases. Contractors to build 3D-printed homes and realtors to sell them are becoming promising careers on the red planet. Lawmakers may also be in great demand as new constitutions and laws will be created to regulate jobs and protect the landscape from human impact. Other occupations may emerge around mining metals and minerals that could be useful for industrial applications. A few large companies from Earth have built factories in space to produce innovative materials and products and export them to Earth.

Eva, an explorer by nature, enjoys the trip more than anyone else in her family. For her, the most fulfilling and exciting part of the trip is gazing at Earth from the spaceship window. She muses to herself that it is "the most beautiful thing I have ever seen and experienced. I feel even more connected with the people on Earth, even though I am thousands and thousands of miles away." What amazes her the most is the stunning, shining "blue marble"—swirls of clouds over the expansive oceans. However, she also identifies alarming sights on the surface, mainly due to climate change. Shrinking ice sheets are noticeable. Coastlines have been reshaped by rising sea levels, transforming large cities like Miami, New York, Guangzhou, and Mumbai. This trip served as a motivation for Eva for two reasons: she wanted to experience the progress made by humans on Mars, but it's also a reminder that important actions still need to be taken on Earth to mitigate impacts from climate change. Many space researchers, scientists, engineers, and new scientific advances have inspired Eva. She now wants to replicate their pioneer spirit to explore more creative ways to use technology to boost humanity's progress and sustainability on Planet Earth.

Regardless of the speed and severity of disruptions created by new technologies, the traits that **make us uniquely human** will remain our key differentiators: **perseverance, adaptability, relationship-building, communication, critical thinking, and integrity.**

Epilogue
Chapter One

Intending to build a future full of new possibilities for ourselves and society, I have decided to call the last section of my book Chapter One. In my view, to adapt to the fast-arriving technological revolutions that will shape our personal lives and the world over the next three decades, we will need to embrace lifelong learning and continually reinvent ourselves. Therefore, instead of final thoughts or conclusions, I want **2050: The Singular Human Era** to become the beginning of a bright and optimistic future. The information, scenarios, and insights in this book are meant to help you imagine and create new possibilities for your personal and professional lives. We can all play a role in building a more positive, sustainable, and equal world and position ourselves at the forefront of change to control our own lives and destinies.

Lifelong learning, upskilling, and self-reinvention will be critical for us to thrive in the twenty-first century and possibly beyond. By focusing on these three key areas, we will be ready to adapt to any unforeseen scenarios created by the next waves of change—technological disruption, shifting demographics, climate change, and inequalities. Changes will be more frequent and severe, so devoting energy and commitment to our personal growth by cultivating digital and human skills will be vital for sustaining our distinctiveness and uniqueness in the future. Moreover, today's disruptions and transformations, driven by exponential technologies, are a preview of the nearly unrecognizable future we could see by 2050.

Technology has been vital to transforming the world into a better place. In the coming years, we must continue to build and leverage new technologies to tackle the world's most pressing issues and mitigate gaps in many areas that require urgent attention, including education, healthcare, digital divide, discrimination, and income and gender inequalities. The exponential growth of AI, cybersecurity, quantum, cloud computing, data science, and other technologies need to improve our lives and sharpen our skills instead of replacing our human abilities, autonomy, and relationships entirely.

The Answer: **Continual Learning**

The future will belong to those who prioritize and practice continual learning. The most fearless and motivated individuals, who incessantly seek new possibilities in a chaotic world, will serve as the impetus for humanity's progress in the twenty-first century. The journey of a lifelong learner contains infinite possibilities, as every day, in every situation, and at every opportunity, they pursue new knowledge and experiences.

As human beings living in the digital era, we should not limit our potential, knowledge, or dreams. Our skills and abilities will likely be irrelevant in a few years; as a singular human, you cannot wait until challenges arise. Therefore, early preparation, discipline, and consistency will be vital to equipping yourself with the essential skills to thrive in the twenty-first century.

The Technological **Revolution**

In 2050 I will be sixty-two years old—the same age as my father today. My son will be thirty-three years old, which is my age today. Although my father and I have experienced accelerated technological advances in our personal and professional lives, the innovations that my son and his generation will experience over the next three decades will be disproportionate to anything we have witnessed before.

In the late 1980s, my parents were in their thirties. It is strange to imagine them relying on a floppy disk to store and share data. Email and instant messaging services were not available yet; communications had to be written or printed on paper, sealed in an envelope, and delivered via human messengers. The introduction of personal computers and software helped my father handle and organize the complex array of data required to manage his projects. Still, before this was an option, my father had to rely on a combination of many manual tools to complete his engineering projects and calculations, like compasses, set squares, and rulers. Instructors conducted training and lectures in person and used physical books or manuals to teach people, while overhead projectors featuring summarized content were standard fixtures in most conference rooms. Remote work was nearly prohibited, as access to the technology involved high costs, and most types of equipment were hardwired, like desktop computers or landline telephones. Recording a birthday party required a big video camera and a couple of film rolls—nearly seven pounds of technology rested on my parent's shoulders to capture these life moments. Listening to music required physical discs; thanks to the first Walkman, they could listen to one or two different collections while they were moving.

Since this time, technology has evolved significantly. It's hard to believe how many devices have converged into a single, super-fast, and powerful mobile phone. As Peter Diamandis and Steven Kotler wrote in their book *The Future Is Faster Than You Think,* thanks to Moore's law, our everyday mobile phones are a "thousand times smaller, a thousand times cheaper, and a million times more powerful than a supercomputer from the 1970s."

Over the last fifteen years, I have witnessed the emergence and disappearance of many technologies—iPods, Myspace, USB flash drives, DVDs. I can still remember the screeching dial-up modem, the steps to connect to the internet, and the ridiculously long time it took to download videos. Today email is quickly becoming obsolete, replaced by instant messaging platforms and tools that allow us to work faster, collaborate more, and get things done more effectively. People can connect with others globally in real-time using videoconferencing and collaborating tools. In addition, AI and machine learning help us create more engaging meetings and webinars, as they help us capture tasks and create meeting minutes.

Mobility is now a key differentiator in attracting and retaining talent; companies are currently rethinking their operating models to build a more connected and engaged workforce by allowing their employees to work from anywhere, at any time. Information and data are growing at an exponential rate. Instead of using physical storage like floppy disks, CDs, or external hard drives, I can easily upload files to cloud platforms so anyone on my team can securely exchange, collaborate on, access, and share working data virtually and instantaneously. Wearable devices can track every step I take during the day, measuring my heartbeat and coaching me to slow down during exercise routines. Always-on algorithms suggest new restaurants, movies, songs, and friends—and all of this information is available in the palm of my hand.

Over the next thirty years, as we experience more technological changes than the previous two and a half centuries, we will need to remain flexible to adapt to the ever-changing world. We must be aware of technological disruptions and transformations to create new and more meaningful possibilities for our lives in a digital future.

Looking Forward **to 2050**

When my son's generation reaches their thirties by the middle of the twenty-first century, the world will be almost unrecognizable. By 2050 communities, industries, and jobs will be entirely reinvented. The world will be hyperconnected by cutting-edge technologies, including AI, virtual reality, quantum computing, blockchain, cloud computing, biotechnology, 3D printing, and many others that will provide solutions for nearly ten billion people living on Planet Earth. As the lines between physical, digital, and biological realms blur, his generation needs to be ready to respond to disruptions and discontinuity driven by digital technologies. Algorithms and social media platforms will better understand people's motivations, moods, and preferences. Younger generations need to realize digital technologies and virtual spaces can be beneficial and detrimental. To thrive in the digital era, they will constantly need to balance their hybrid realities. They hold the key to a better future; therefore, they need to learn about digital ethics, develop human skills, respect different perspectives, and become global thinkers to navigate a dynamic, unpredictable, and diverse world.

Work will be radically reconfigured as jobs are equally divided between humans and machines. The most exciting jobs will be those we cannot yet predict. As machines and automation take the most repetitive, monotonous, heavy-lifting work away from humans, today's concept of work will need to be redefined. AI, automation, and robots will help people access more products and build more connections, but they will also reshape the relationship between salaries and work. The increased presence of machines in the workplace will eliminate the need for humans to perform repetitive and routine tasks. The rapid advance of automation will require people to learn new skills to ensure they have alternatives when it's time to transition from one occupation to another. Our worldwide perspectives, behaviors, and skills will need to be recycled, reimagined, and reinvented at the same speed and scale as the changes around us, so we aren't displaced or rendered obsolete.

Personal computers will likely vanish, and eventually, virtual realities and brain-computer interfaces will displace smartphones as people use their eyes or brains to connect directly to the digital world. The evolution from screens to virtual reality and augmented realities to brain-computer interface technologies will challenge and change every aspect of our lives, including education, entertainment, communication, healthcare, and how we interact with each other and with machines. An endless amount of content will be seamlessly available anywhere, anytime, without any need for search engines. Devices and sensors in and outside of our bodies powered by AI and machine learning algorithms will be able to recommend movies, songs, and products based on our moods and preferences as they get better at understanding human thoughts and emotions. As the lines between reality and virtuality fade, we must stay mindful that the more machines can access and understand our human behaviors, intuitions, and decisions, the more they can influence and persuade us.

Healthcare will progress astronomically, shifting to precise, preventive, and personalized healthcare, resulting in a significant extension of the human lifespan and healthspan. We are already exploring the human genome by tracing and identifying the genes that may hold certain chronic diseases, allergies, or food intolerances. As biotechnology evolves, its enhanced capabilities will enable us to "correct" affected genes safely, hopefully preventing or curing severe illnesses like cancer and heart disease. Advances in nanotechnology could send tiny computer chips into the human body to perform cellular repairs or deliver drugs exactly where they're needed. Smartwatches already track our heartbeats and blood pressure. Soon, new devices such as smart clothes will help detect early signs of a heart attack and identify any missing nutrients in our daily diet. In the coming decades, we will see more centenarians in many parts of the world. The oldest people will be either the rich, as they can afford to extend their lives, or the individuals who practice healthy habits and manage high-stress levels.

Although some of these technological advancements seem ambitious, there are indicators they will be realistic options in the future. Of course, some developments bring uncertainty and raise ethical concerns. Still, I'm optimistic that groundbreaking technologies will continue to create new and positive possibilities for humanity in the twenty-first century. They have the potential to solve the most critical world challenges.

What Will **Not Change**

Understanding what will not change by 2050 is equally important as making sense of what will change. Doing so will give us an advantage in an uncertain and unpredictable future. Regardless of the speed and severity of disruptions created by new technologies, the traits that make us uniquely human will remain our key differentiators: perseverance, adaptability, relationship-building, communication, critical thinking, and integrity. These cannot be automated or digitized, and thus they will become scarcer and more valuable in the future. If we focus on becoming adept at these human capabilities, we will continue to add value to our lives and the companies we engage with. Additionally, these human capabilities will prepare us to adapt to changes and uncertain conditions in the future.

Technological innovations have always served to empower and help people, and they will continue to be catalysts in humanity's progress. A machine working without human cooperation and oversight will not be more powerful than a machine actively working alongside a human being. Human doctors will update and augment their capabilities by leveraging precise diagnoses from AI systems, but an AI physician or a nurse robot will not replace them. Patient relationships will still demand sensitivity, empathy, and compassion from another human, mainly when dealing with emotional circumstances and mental health conditions. Likewise, engineers and designers can leverage digital tools, AI, and big data to help them create innovative projects. Still, humility, sociability, and social awareness are vital to building more intelligent and sustainable cities. Learning will be highly personalized, thanks to AI capabilities, cloud computing, chatbots, and virtual reality; these will transform teachers into content moderators while saving them time to focus on inspiring, coaching, and mentoring students. Software development and engineering will require greater human oversight to tackle data bias and ethical issues. Highly sophisticated technology, scientific research, robots, and spaceships will allow further discoveries on the moon and Mars. Still, they would not be possible without the insatiable human curiosity to understand our origins, evolution, and surroundings.

Everything around us will change in some form; the way we live, work, receive healthcare, educate ourselves, and interact with others will be transformed due to breakthrough technologies around us. Humanity will rely on explorers, visionaries, and risk-takers to create innovations and solutions to solve global challenges and improve people's well-being. The world will continue to reward courageous individuals who are intentional about their future. In the twenty-first century, the people who can continually reinvent themselves, learn from failures, replicate successes, and embrace change are the ones who will succeed, thrive, and position themselves to build a better future for themselves and humanity at large. Commitment to self-improvement was vital yesterday, and it will remain a key differentiator tomorrow.

As we digitize many aspects of our lives, including work, entertainment, education, and shopping, we must constantly remind ourselves that protecting our human values and nurturing human-to-human relationships are our top priorities. We need to keep our family and friends close to our hearts; cultivating relationships with our loved ones will be more critical than ever, especially as we navigate times of increased uncertainty. Machines may be by our sides every minute of every day, tracking our steps, actions, and behaviors, but human interaction will remain irreplaceable.

As AI, automation, and digitization quickly expand to new territories, many existing jobs will be eliminated or completely redesigned in the coming years. However, what will not change is the need for highly adaptable, trustworthy, collaborative, problem-solving, and creative human workers who are willing to learn, unlearn, and relearn skills. Companies will increasingly need a pool of diverse and forward-thinking people who can genuinely understand humanness and translate that unique quality into more personalized, sustainable, and inclusive products to create an environment suitable for all people and the planet. In addition, the complexity and speed of technological advances, new business models, and always-changing customer needs will require increased human involvement, oversight, intervention, and decision-making. As a result, organizations of all sizes will need people's brainpower, flexibility, imagination, intuition, creativity, and critical thinking in more intricate ways across different domains to solve global-scale problems. While AI and machine learning crunch data and provide more precise calculations at speed beyond human ability, traits such as reasoning, advising, negotiating, resolving conflict, and translating ideas to different scenarios will be unique human attributes.

We need to embrace challenges and remain flexible. We need to look at changes openly and holistically and find new ways to turn uncertainty and adversity into new possibilities. New challenges will always keep coming our way. Still, our ability to recognize and understand how the next waves of change might impact our lives and the world allow us to anticipate and develop the skills to thrive in an always-changing world. The desire to continue moving forward, exploring, and expanding human skills to meet current and long-term needs will remain an invaluable differentiator for singular humans. Natural selection in the twenty-first century will favor humans who can imagine alternatives, define a clear path toward a more fulfilling future, and constantly reinvent themselves to flourish in an unpredictable world. As what is relevant today might quickly become obsolete tomorrow, people need to practice the Six C's habit. You need to be **creative** to find new ideas and opportunities to achieve your full potential as a human being. Have the **courage** to embrace new challenges and scenarios, and keep reinventing yourself. **Commit** your time to projects, jobs, and dreams that will provide you with a sense of purpose, fulfillment, and joy. Be **consistent** with your self-development, learn new skills continually, and remain flexible to adapt to new situations. **Conquer** obstacles, uncertainties, and fears; you need to grit in the face of difficulties to achieve your long-term goals. Finally, **celebrate** your victories and achievements as a daily habit.

The Opportunities **Will Be Incredible**

We live in one of the most extraordinary times of humanity's history. The opportunities for people to do what they genuinely enjoy and add value for themselves and the world are limitless. The scenarios in this book are not meant to predict the future; instead, those insights serve as tools to help you think and make sense of the changes that are redefining our lives and the world. With this book, my goal is to provide trusted information so you can imagine and immerse yourself in future scenarios, consider what human skills you need to develop to make your dreams a reality, and create a life of possibility. This book aims to achieve continuity; just like learning, it is continuous and life-spanning. The world is changing faster than ever; over the next thirty years, everything we know now will be reinvented, recycled, reused, or relearned at an unprecedented speed. It would be premature to predict every transformation and necessary skill to access jobs yet to be created. In the spirit of continuity, this is a living book—dynamic, endless, and everlasting. I believe that learning is infinite, and knowledge, experiences, discoveries, and life stories are meant to be shared and celebrated with one another. Collaboration, communication, critical thinking, open dialogues, and respecting diverse perspectives are vital tools to build a cohesive society.

Looking ahead, I aim to continue providing compelling content and learning experiences through my channel so you can stay aware and learn about the key trends shaping our lives, industries, and the world. The winners of the twenty-first century will be lifelong learners and explorers open to new and unforeseen scenarios. Conversely, the people unwilling to embrace changes and failing to prepare themselves to access the jobs of tomorrow will be left behind.

I will continue to study, explore, learn, and share relevant insights and foresight about breakthrough innovations that will disrupt the world in the years ahead and how we can adapt to always-changing scenarios. I am committed to helping you build confidence and new possibilities through continual learning so you can sharpen the uniquely human skills that will accelerate your success and distinguish you in the twenty-first century. I invite you to stay connected with me by discussing, debating, co-creating, and sharing ideas for us to create a better future, and most importantly, to learn new topics on my informational hub: **www.rafaelirio.com.** My goal is to transform this book into the most valuable and human-centered playbook in the twenty-first century so we can use it to thrive in a future defined by rapid advances in technology.

I have three recommendations for the following chapters of your life: start exploring new possibilities today, embrace challenges and change, and frequently reinvent yourself so you can continue growing personally and professionally. Change is the only certainty in an unpredictable future.

Acknowledgments

Since 1988 the most important teachers of my life—my parents, Luiz Carlos and Rosana—have instilled in me the most valuable life lesson: the art of possibility. They have encouraged me to explore the unknown and experience new ideas, countries, and cultures. They helped me become a global citizen passionate about finding new solutions for real-world issues. Since my early days, they have constantly reminded me that we are responsible for building our future; and to live a fulfilling and purposeful life, we need to persevere and take responsibility for our actions and their outcomes. My parents taught me that we could continue to grow personally and professionally by embracing change and new challenges.

Most importantly, they taught me to believe that each new day will be better than the last. They have shown me that anything and everything is possible when we have big dreams, a future-ready mindset, and a willingness to take risks. I am forever grateful to them for this legacy.

I could not have finished writing this book without the unwavering support of my wife, Laís, and my son, Pedro. These two human beings have inspired me to reach the finish line. I had been dreaming about this project for years—it has become a reality thanks to their comprehension, empathy, patience, and unconditional love. "When I finish my book," or "hopefully, I will next month," are examples of the responses they received to so many invitations throughout last year—their understanding and acceptance of the circumstances were invaluable.

I am grateful to all my other family members and friends for their assistance and companionship. Coming from Brazil, the home of more than 200 million creative citizens, I was taught core values such as integrity, courage, authenticity, accountability, and gratitude. These have been invaluable to me as the foundation of myself as a human being, serving as my daily compass and helping me to believe that diversity is humanity's biggest asset. To all the colleagues I had the pleasure to meet in every corner of the world during the last decade, your stories, ideas, and guidance have inspired me further to explore the intersection between technology and human beings.

About the Author

The most appropriate word to define my life is **change**. Explorations, discoveries, adaptability, and perseverance represented my personal and professional life in the last decade. I have spent the last ten years working in the Information Technology (IT) sector for one of the largest global technology companies, which serves 75 of the Top Fortune 100 companies. This opportunity has allowed me to expand my career into international markets across Latin America, Asia Pacific, and North America. The chance to live in four countries has enhanced me professionally, personally, and culturally. By immersing myself in different cultures, I have learned, embraced, and shared the importance of core values such as accountability, diversity, fairness, integrity, and initiative. These values have shaped my beliefs, perspectives, and actions, making me a global citizen prepared to help people navigate an always-changing world.

In 2013, I embarked on one of my life's most exciting journeys. I moved to Asia. I lived in Shanghai, China, and Singapore for three years altogether. I oversaw sales and business operations for around 150 countries worldwide during those years. My time in Asia was a unique journey into the unknown. Every day was a new opportunity to discover, explore, and co-create new possibilities with my global and diverse team.

I landed in the United States in 2015. Since then, I have held multiple positions within the IT sector, such as leading operations and business transformations for a worldwide technology group and managing sales and technical programs for a large infrastructure services organization in North America. Moving to the United States, home of the most innovative companies in the world, made me believe and advocate that digital technologies can serve as a vehicle to improve people's well-being and augment their capabilities and intelligence.

My personal and professional experiences have accumulated into one core message: human beings are and will continue to be the cornerstone of humanity's progress. Regardless of our nationality, race, gender, or sexual preferences, we are all visionaries, forward-thinking individuals, and architects of our own lives. However, to create and enjoy the life we desire, we must constantly remind ourselves that the future is ours to build. I believe that we – each one of us – can change the world for the better, taking tangible and consistent actions to create new possibilities that will benefit ourselves and society as a whole. Conversely, being paralyzed by the uncertainty of an unpredictable future will likely lead some people to obsolescence. Therefore, my recommendation for anyone looking for a better future is to start exploring new possibilities today, embrace challenges and changes, and frequently reinvent yourself so you can continue growing personally and professionally.

My purpose is to inspire and empower all explorers and lifelong learners worldwide to expand their human potential in a world pressured by accelerated technological advances.

rafael IRIO

References

Front Matter:

Diamandis, Peter H. *"Proof of Exponential Tech Growth."* Blog post.
https://www.diamandis.com/blog/proof-exponential-tech-growth.
Accessed November 12, 2021.

"Human being." Lexico Dictionaries.
https://www.lexico.com/en/definition/human_being. Accessed November 12, 2021.

International Data Corporation. *"AR & VR Headsets Market Share."*
https://www.idc.com/promo/arvr.
Accessed December 23, 2021.

United Nations. *"Our World Is Growing Older: UN DESA Releases New Report on Ageing."*
https://www.un.org/development/desa/en/news/population/our-world-is-growing-older.html.
Accessed November 12, 2021.

World Economic Forum. *"Emerging Economies More Optimistic about Artificial Intelligence."*
January 05, 2022.
https://www.weforum.org/press/2022/01/emerging-economies-more-optimistic-about-artificial-intelligence-survey-finds.
Accessed January 13, 2022.

Worldometers. *"World Population: Past, Present, and Future."*
https://www.worldometers.info/world-population/.
Accessed December 12, 2021.

Chapter 1:

AeroFarms. *"Growing in Vertical Farms for Elevated Flavor."*
https://www.aerofarms.com/farms/.
Accessed November 12, 2021.

AFN. *"Food Waste Startup Hazel Technologies Raises Oversubscribed $13m Series B to Scale Up Production."* August 12, 2019.
https://agfundernews.com/food-waste-startup-hazel-technologies-raises-oversubscribed-13m-series-b-to-scale-up-production.html.

BMW. *"Urban Mobility: The 5 Trends of the Future."*
https://www.bmw.com/en/innovation/5-trends-of-urban-mobility.amp.html.
Accessed November 12, 2021.

Bradberry, Travis, and Jean Greaves. *Emotional Intelligence 2.0. TalentSmart*, 2009.

C40 Cities. *"A Global Network of Mayors Taking Urgent Action to Confront the Climate Crisis and Create a Future Where Everyone Can Thrive."*
https://www.c40.org/.
Accessed November 12, 2021.

———. *"Sea Level Rise and Coastal Flooding."*
https://www.c40.org/other/the-future-we-don-t-want-staying-afloat-the-urban-response-to-sea-level-rise.
Accessed November 12, 2021.

California Department of Justice, Office of the Attorney General. *"California Consumer Privacy Act (CCPA)."* October 15, 2018.
https://oag.ca.gov/privacy/ccpa.

Campbell, Sophia, Jimena Ruiz Castro, and David Wessel. *"The Benefits and Costs of Broadband Expansion."* Brookings Institution, August 18, 2021.
https://www.brookings.edu/blog/up-front/2021/08/18/the-benefits-and-costs-of-broadband-expansion/.

Credit Suisse. *"The Global Food System: Identifying Sustainable Solutions."* 2021. https://www.credit-suisse.com/media/assets/corporate/docs/about-us/research/publications/the-global-food-system-identifying-sustainable-solutions.pdf.

CriticalThinking.org. *"Defining Critical Thinking."*
https://www.criticalthinking.org/pages/defining-critical-thinking/766.
Accessed November 12, 2021.

Diamandis, Peter H. *"Neuralink . . . An Artificial Intelligence-Mind Merger."* Blog post.
https://www.diamandis.com/blog/neuralink-ai-mind-merger. Accessed November 12, 2021.

Diamandis, Peter H., and Steven Kotler. *The Future Is Faster Than You Think.* Simon & Schuster, 2020.

E-Estonia. *"We Have Built a Digital Society and We Can Show You How."*
https://e-estonia.com/wp-content/uploads/e-estonia-200121-eng.pdf
Accessed November 12, 2021.

"The Farms Being Run from Space." BBC.
https://www.bbc.com/future/bespoke/follow-the-food/the-farms-being-run-from-space/.
Accessed November 12, 2021.

Finless Foods. *"About."*
https://finlessfoods.com/about/#products.
Accessed November 12, 2021.

Food and Agriculture Organization of the United Nations. *"The State of Food Security and Nutrition in the World 2021."*
http://www.fao.org/3/cb4474en/online/cb4474en.html#chapter-executive_summary. Accessed November 12, 2021.

GDPR. *"What Is the LGPD? Brazil's Version of the GDPR."* July 31, 2019.
https://gdpr.eu/gdpr-vs-lgpd/.

Good Food Institute. *"2020 State of the Industry Report: Cultivated Meat."*
https://gfi.org/wp-content/uploads/2021/04/COR-SOTIR-Cultivated-Meat-2021-0429.pdf.

———. *"The Science of Cultivated Meat."*
https://gfi.org/science/the-science-of-cultivated-meat/. Accessed November 12, 2021.

Hamilton, Isobel Asher. *"The Science of Elon Musk's Neuralink, Which Plans to Put Chips into People's Brains."* Business Insider, August 26, 2020.
https://www.businessinsider.com/we-spoke-to-2-neuroscientists-about-how-exciting-elon-musks-neuralink-really-is-2019-9.

Housing and Development Board. *"Public Housing – A Singapore Icon."*
https://www.hdb.gov.sg/about-us/our-role/public-housing-a-singapore-icon.
Accessed November 12, 2021.

IEA. *"Net Zero by 2050—Analysis."*
https://www.iea.org/reports/net-zero-by-2050.
Accessed November 12, 2021.

IFTF. *"The Hyperconnected World of 2030–2040."*
https://www.iftf.org/hyperconnectedworld/.
Accessed November 12, 2021.

"Industrial Revolution." Encyclopædia Britannica.
https://www.britannica.com/event/Industrial-Revolution.
Accessed November 12, 2021.

"Industrial Revolution and Technology." National Geographic, January 9, 2020.
https://www.nationalgeographic.org/article/industrial-revolution-and-technology/.

Infarm. *"Infarm Heralds the Next Generation of Cloud-Enabled Farming."* https://www.infarm.com/infarm-heralds-the-next-generation-of-cloud-enabled-farming/. Accessed November 12, 2021.

Lewis, Nell, and Samuel Burke. *"How Daily Life Will Benefit from 5G."* CNN, February 28, 2019. https://www.cnn.com/2019/02/28/tech/5g-benefits-mobile-world-congress/index.html.

Marr, Bernard. *Tech Trends in Practice: The 25 Technologies that are Driving the 4th Industrial Revolution.* John Wiley & Sons, 2020.

McKinsey & Company. *"Smart City Technology for a More Liveable Future."* June 5, 2018. https://www.mckinsey.com/business-functions/operations/our-insights/smart-cities-digital-solutions-for-a-more-livable-future.

. *"Cultivated Meat: Out of the Lab, into the Frying Pan."* June 16, 2021. https://www.mckinsey.com/industries/agriculture/our-insights/cultivated-meat-out-of-the-lab-into-the-frying-pan.

———. *"How Cities Can Adapt to Climate Change."* July 20, 2021. https://www.mckinsey.com/business-functions/sustainability/our-insights/how-cities-can-adapt-to-climate-change.

———. *"What the Draft European Union AI Regulations Mean for Business."* August 10, 2021. https://www.mckinsey.com/business-functions/mckinsey-analytics/our-insights/what-the-draft-european-union-ai-regulations-mean-for-business.

Moody, Joanna, Elizabeth Farr, Marisa Papagelis, and David R. Keith. *"The Value of Car Ownership and Use in the United States."* Nature Sustainability 4 (2021): 769–74. https://doi.org/10.1038/s41893-021-00731-5.

Niiler, Eric. *"How the Second Industrial Revolution Changed People's Lives."* History.com. https://www.history.com/news/second-industrial-revolution-advances. Accessed November 12, 2021.

Office of the Director of National Intelligence. *"Global Trends 2040: A More Contested World."* 2021. https://www.dni.gov/files/ODNI/documents/assessments/GlobalTrends_2040.pdf.

Olick, Diana. "3D-Printed Housing Developments Suddenly Take Off. Here's What They Look Like." CNBC, March 12, 2021. https://www.cnbc.com/2021/03/12/3d-printed-housing-developments-suddenly-take-off-heres-what-they-look-like.html.

Our World in Data. *"Land Use."* https://ourworldindata.org/land-use. Accessed November 12, 2021.

———. *"What Are the Environmental Impacts of Food and Agriculture?"* https://ourworldindata.org/env-impacts-of-food. Accessed November 12, 2021.

Palmer, Danny. *"What Is GDPR? Everything You Need to Know about the New General Data*

Protection Regulations." ZDNet, May 17, 2019.
https://www.zdnet.com/article/gdpr-an-executive-guide-to-what-you-need-to-know/.

Plumer, Brad, Nadja Popovich, and Blacki Migliozzi. *"Electric Cars Are Coming. How Long Until They Rule the Road?."* New York Times, March 11, 2021.
https://www.nytimes.com/interactive/2021/03/10/climate/electric-vehicle-fleet-turnover.html.

Recycle Track Systems. *"Food Waste in America in 2021: Statistics & Facts."*
https://www.rts.com/resources/guides/food-waste-america/. Accessed November 12, 2021.

"Revolution." Merriam-Webster.
https://www.merriam-webster.com/dictionary/revolution. Accessed December 29, 2021.

Sample, Ian. *"Universal Internet Access Unlikely until at Least 2050, Experts Say."* Guardian, January 10, 2019.
https://www.theguardian.com/technology/2019/jan/10/universal-internet-access-unlikely-until-2050-experts-say-lack-skills-investment-slow-growth.

Singularity Hub. *"An Israeli Startup Is 3D Printing Cultured Ribeye Steaks."* February 17, 2021.
https://singularityhub.com/2021/02/17/an-israeli-startup-is-3d-printing-cultured-ribeye-steaks/.

Starlink. Home page.
https://www.starlink.com/.
Accessed November 12, 2021.

Statista. *"Internet Users in the World 2021."*
https://www.statista.com/statistics/617136/digital-population-worldwide/.
Accessed November 12, 2021.

Statista. *"Starlink Satellite Numbers 2021."*
https://www.statista.com/statistics/1224164/starlink-satellite-launches/.
Accessed November 12, 2021.

Tsai, Katie. *"Cell-Based Seafood Producer BlueNalu Raises $60 Million."* CNBC, January 19, 2021.
https://www.cnbc.com/2021/01/19/cell-based-seafood-producer-bluenalu-raises-60-million.html.

United Nations. *"Feeding the World Sustainably."*
https://www.un.org/en/chronicle/article/feeding-world-sustainably.
Accessed November 12, 2021.

———. *"Goal 11: Make Cities Inclusive, Safe, Resilient and Sustainable."*
https://www.un.org/sustainabledevelopment/cities/.
Accessed November 12, 2021.

———. *"Make Cities and Human Settlements Inclusive, Safe, Resilient and Sustainable."*
https://unstats.un.org/sdgs/report/2019/goal-11/.
Accessed November 12, 2021.

———. *"Sustainable Consumption and Production."*

https://www.un.org/sustainabledevelopment/sustainable-consumption-production/.
Accessed November 12, 2021.

———. *"68% of the World Population Projected to Live in Urban Areas by 2050, Says UN."* May 16, 2018.
https://www.un.org/development/desa/en/news/population/2018-revision-of-world-urbanization-prospects.html.
Accessed November 12, 2021.

———. *"Internet Milestone Reached As More Than 50 per Cent Go Online."* UN News, December 7, 2018.
http://news.un.org/en/story/2018/12/1027991.

United States Energy Information Administration. Home page.
https://www.eia.gov/outlooks/ieo/pdf/IEO2021_Narrative.pdf.
Accessed November 12, 2021.

Versace, Chris, Lenore Elle Hawkins, and Mark Abssy. *"World Reimagined: The Future of Plant-Based Foods."* Nasdaq, December 9, 2020.
https://www.nasdaq.com/articles/world-reimagined%3A-the-future-of-plant-based-foods-2020-12-09.

Wikipedia. *"Mobility As a Service."* Last updated October 10, 2015.
https://en.wikipedia.org/wiki/Mobility_as_a_service.

World Economic Forum. *"The Fourth Industrial Revolution: What It Means and How to Respond."* January 2016.
https://www.weforum.org/agenda/2016/01/the-fourth-industrial-revolution-what-it-means-and-how-to-respond/.
Accessed November 12, 2021.

World Economic Forum. *"What Is Vertical Farming? And What Are the Benefits?"* September 2019.
https://www.weforum.org/agenda/2019/09/vertical-farming-agriculture-aerofarms/.
Accessed November 12, 2021.

World Economic Forum. *"When Will We Be Eating Lab-Grown Meat?"* October 2020. https://www.weforum.org/agenda/2020/10/will-we-eat-lab-grown-meat-world-food-day/.
Accessed November 12, 2021.

World Health Organization. *"Obesity and Overweight."*
https://www.who.int/news-room/fact-sheets/detail/obesity-and-overweight.
Accessed November 12, 2021.

Worldometers. *"World Population: Past, Present, and Future."*
https://www.worldometers.info/world-population/.
Accessed November December 12, 2021.

World Resources Institute. *"6 Pressing Questions about Beef and Climate Change, Answered."*
https://www.wri.org/insights/6-pressing-questions-about-beef-and-climate-change-answered.
Accessed November 12, 2021.

Zhao, Xiaoying. *"Forget 5G, the US and China Are Already Fighting for 6G Dominance."* Bloomberg, February 8, 2021.
https://www.bloomberg.com/news/features/2021-02-08/forget-5g-the-u-s-and-china-are-already-fighting-for-6g-dominance.
Accessed November 12, 2021.

Chapter 2:

Bloomberg. *"Batteries For Electric Cars Speed Toward a Tipping Point."*
https://www.bloomberg.com/news/articles/2020-12-16/electric-cars-are-about-to-be-as-cheap-as-gas-powered-models.
Accessed December 27, 2021

Briggs, Helen. *"Biodiversity loss risks 'ecological meltdown'"* BBC, October 10, 2021. https://www.bbc.com/news/science-environment-58859105.

Credit Suisse. *"The Global Food System: Identifying Sustainable Solutions."* 2021. https://www.credit-suisse.com/media/assets/corporate/docs/about-us/research/publications/the-global-food-system-identifying-sustainable-solutions.pdf.

———. *"Why Wealth Matters. The Global Wealth Report."*
https://www.credit-suisse.com/about-us/en/reports-research/global-wealth-report.html.

Defense Advanced Research Projects Agency. *"Teaching Machines Common Sense Reasoning."* https://www.darpa.mil/news-events/2018-10-11. Accessed December 27, 2021.

Diamandis, Peter H., and Steven Kotler. *The Future Is Faster Than You Think. Simon & Schuster,* 2020

Eyesafe. *"COVID-19: Screen Time Spikes to over 13 Hours per Day According to Eyesafe Nielsen Estimates."* March 28, 2020.
https://eyesafe.com/covid-19-screen-time-spike-to-over-13-hours-per-day/.

Intel. *"Over 50 Years of Moore's Law."*
https://www.intel.com/content/www/us/en/silicon-innovations/moores-law-technology.html.
Accessed November 12, 2021.

Investopedia. *"Moore's Law Definition."* November 24, 2003.
https://www.investopedia.com/terms/m/mooreslaw.asp.

McKinsey & Company. *"How COVID-19 has pushed companies over the technology tipping point—and transformed business forever."* October 5, 2020.
https://www.mckinsey.com/business-functions/strategy-and-corporate-finance/our-insights/how-covid-19-has-pushed-companies-over-the-technology-tipping-point-and-transformed-business-forever.

———. *"Internet Matters: The Net's Sweeping Impact on Growth, Jobs, and Prosperity."* May 1, 2011.
https://www.mckinsey.com/~/media/McKinsey/Industries/Technology%20Media%20and%20Telecommunications/High%20Tech/Our%20Insights/Internet%20matters/MGI_internet_matters_exec_summary.pdf.

———. *"The top trends in tech—executive summary download."* June 24, 2021. https://www.mckinsey.com/~/media/mckinsey/Business%20Functions/McKinsey%20Digital/Our%20Insights/The%20top%20trends%20in%20tech%20final/Top-trends-in-tech-executive-summary-6-24-21.

Marr, Bernard. Tech Trends in Practice: The 25 Technologies that are Driving the 4th Industrial Revolution. John Wiley & Sons, 2020.

McGrath, Matt. *"Climate Change: IPCC Report Is 'Code Red for Humanity.'"* BBC, August 9, 2021.
https://www.bbc.com/news/science-environment-58130705.

NCTA – The Internet & Television Association. *"Broadband Facts & Stats."*
https://www.ncta.com/broadband-facts. Accessed January 4, 2022.

National Human Genome Research Institute. *"The Cost of Sequencing a Human Genome."*
https://www.genome.gov/about-genomics/fact-sheets/Sequencing-Human-Genome-cost. Accessed November 12, 2021.

Neuralink. Home page.
https://neuralink.com/. Accessed November 12, 2021.

Office of the Director of National Intelligence. *"The Future of Migration."*
https://www.dni.gov/index.php/gt2040-home/gt2040-deeper-looks/future-of-migration.

———. *"Global Trends 2040: A More Contested World."* 2021.
https://www.dni.gov/files/ODNI/documents/assessments/GlobalTrends_2040.pdf.

Our World in Data. *"The price of batteries has declined by 97% in the last three decades."*
https://ourworldindata.org/battery-price-decline.
Accessed November 12, 2021.

———. *"Technological Progress."*
https://ourworldindata.org/technological-progress.
Accessed November 12, 2021.

Singularity Hub. *"Microsoft's Mesh Will Let Us Beam Ourselves to Work as Holograms."* March 10, 2021.
https://singularityhub.com/2021/03/10/microsoft-mesh-will-let-us-beam-ourselves-to-work-as-holograms/.

Statista. *"Forecast of the Total Population of Africa from 2020 to 2050."*
https://www.statista.com/statistics/1224205/forecast-of-the-total-population-of-africa/.

———. *"Number of internet users worldwide from 2005 to 2021."*
https://www.statista.com/statistics/273018/number-of-internet-users-worldwide/.
Accessed November 12, 2021.

———. *"Population Projections for the United States from 2015 to 2060."*
https://www.statista.com/statistics/183481/united-states-population-projection/.

The New York Times. *"Are We in the Metaverse Yet?"*
https://www.nytimes.com/2021/07/10/style/metaverse-virtual-worlds.html.

United Nations. *"Report of the UN Economist Network for the UN 75th Anniversary: Shaping the Trends of Our Future."* September 2020.
https://www.un.org/development/desa/publications/wp-content/uploads/sites/10/2020/09/20-124-UNEN-75Report-2-1.pdf.

United Nations Development Programme. *"Wide Inequalities in People's Well-Being Cast a Shadow on Sustained Human Development Progress."*
http://hdr.undp.org/en/content/wide-inequalities-peoples-well-being-cast-shadow-sustained-human-development-progress.

United States Bureau of Labor Statistics. *"Employment in STEM Occupations."*
https://www.bls.gov/emp/tables/stem-employment.htm.

World Economic Forum. *"In the Race to 50 Million Users There's One Clear Winner—and It Might Surprise You."*
https://www.weforum.org/agenda/2018/06/how-long-does-it-take-to-hit-50-million-users. Accessed November 12, 2021.

———. *"Climate talk needs a reset. Here's why."* September 22, 2020.
https://www.weforum.org/agenda/2021/09/climate-talk-needs-a-reset-here-is-why/. Accessed November 12, 2021.

———. *"Global Gender Gap Report 2021: Insight Report."*
http://www3.weforum.org/docs/WEF_GGGR_2021.pdf.

Chapter 3:

Alberti, Fay Bound. *"Robots to Be Introduced in UK Care Homes to Allay Loneliness—That's Inhuman."* Conversation, September 14, 2020.
https://theconversation.com/robots-to-be-introduced-in-uk-care-homes-to-allay-loneliness-thats-inhuman-145879.

American Society of Mechanical Engineers. *"10 Humanoid Robots of 2020."*
https://www.asme.org/topics-resources/content/10-humanoid-robots-of-2020. Accessed November 12, 2021.

Bradberry, Travis, and Greaves, Jean. *Emotional Intelligence 2.0. TalentSmart,* 2009.

"Chat*bot Market 2021: Stats, Trends, Size & Ecosystem Research."* Business Insider, February 8, 2021.
https://www.businessinsider.com/chatbot-market-stats-trends.

Clifford, Cat. *"Elon Musk's Neuralink: Company Has Bold Ideas, Hurdles."* CNBC, December 5, 2020.
https://www.cnbc.com/2020/12/05/elon-musks-neuralink-bold-ideas-hurdles.html.

David, Susan. *Emotional Agility.* Penguin Random House, 2016.

Drew, Liam. *"The Ethics of Brain–Computer Interfaces."* Nature, July 24, 2019. https://www.nature.com/articles/d41586-019-02214-2.

"Elon Musk to Show Off Working Brain-Hacking Device." BBC, August 28, 2020. https://www.bbc.com/news/technology-53921596.

Epstein, David. Range: *Why Generalists Triumph in a Specialized World.* Riverhead, 2019.

Facebook AI. *"A State-of-the-Art Open Source Chatbot."* Blog post. https://ai.facebook.com/blog/state-of-the-art-open-source-chatbot/. Accessed November 12, 2021.

Godin, Seth. *The Dip.* Penguin Publishing Group, 2007.

Google AI. *"Towards a Conversational Agent That Can Chat about . . . Anything."* Blog post, January 28, 2020. https://ai.googleblog.com/2020/01/towards-conversational-agent-that-can.html. Accessed November 12, 2021.

Hao, Karen. *"This Robot Taught Itself to Walk Entirely on Its Own."* MIT Technology Review, March 2, 2020. https://www.technologyreview.com/2020/03/02/905593/ai-robot-learns-to-walk-autonomously-reinforcement-learning/.

Harari, Yuval Noah. *21 Lessons for the 21st Century.* Random House, 2019.

Heaven, Will Douglas. *"Google Says Its New Chatbot Meena Is the Best in the World."* MIT Technology Review, January 30, 2020. https://www.technologyreview.com/2020/01/30/275995/google-says-its-new-chatbot-meena-is-the-best-in-the-world/.

Holley, Peter. *"'Colossus,' a 1,100-Pound Robot, Helped Extinguish Notre Dame Fire in Hard-to-Reach Spots."* Washington Post, April 17, 2019. https://www.washingtonpost.com/technology/2019/04/17/firefighters-had-secret-weapon-when-notre-dame-caught-fire-robot-named-colossus/.

Hutson, Matthew. *"8 Truths about Intuition."* Psychology Today, January 2020. https://www.psychologytoday.com/us/articles/201912/8-truths-about-intuition. Accessed November 12, 2021.

Kharpal, Arjun. *"Stephen Hawking Says AI Could Be 'Worst Event' in Civilization."* CNBC, November 6, 2017. https://www.cnbc.com/2017/11/06/stephen-hawking-ai-could-be-worst-event-in-civilization.html.

Leonhard, Gerd. *"Five New Human Rights for the Digital Age."* In Technology vs. Humanity: The Coming Clash between Man and Machine. Fast Future, 2016. https://www.futuristgerd.com/2017/03/five-new-human-rights-for-the-digital-age-excerpt-from-gerd-leonhards-book-technology-vs-humanity. Accessed November 12, 2021.

Maynard, Andrew. *"Elon Musk's Tesla Bot Raises Serious Concerns—But Probably Not the Ones You Think."* Conversation, September 7, 2021.
https://theconversation.com/elon-musks-tesla-bot-raises-serious-concerns-but-probably-not-the-ones-you-think-166714.

McKinsey & Company. *"The Most Fundamental Skill: Intentional Learning and the Career Advantage."* August 7, 2020.
https://www.mckinsey.com/featured-insights/future-of-work/the-most-fundamental-skill-intentional-learning-and-the-career-advantage.

Oberlo. *"10 Artificial Intelligence Statistics You Need to Know in 2021."* August 22, 2020.
https://www.oberlo.com/blog/artificial-intelligence-statistics.

Prescott, Tony. *"Will Robots Make Good Friends? Scientists Are Already Starting to Find Out."* Conversation, February 15, 2021.
https://theconversation.com/will-robots-make-good-friends-scientists-are-already-starting-to-find-out-154034.

Reisinger, Don. *"All Companies Should Live by the Jeff Bezos 70 Percent Rule."* Inc.
https://www.inc.com/don-reisinger/all-companies-should-live-by-jeff-bezos-70-percent-rule.html.
Accessed November 12, 2021.

Romero, María Elena. *"Tommy the Robot Nurse Helps Italian Doctors Care for COVID-19 Patients."* Public Radio International.
https://www.pri.org/stories/2020-04-08/tommy-robot-nurse-helps-italian-doctors-care-covid-19-patients.
Accessed November 12, 2021.

Shead, Sam. *"Elon Musk Says DeepMind Is His 'Top Concern' When It Comes to A.I."* CNBC, July 29, 2020.
https://www.cnbc.com/2020/07/29/elon-musk-deepmind-ai.html.

Voicebot.Ai. *"Google's New Meena Chatbot Imitates Human Conversation and Bad Jokes."* February 3, 2020.
https://voicebot.ai/2020/02/03/googles-new-meena-chatbot-imitates-human-conversation-and-bad-jokes/.

"What Having a 'Growth Mindset' Actually Means." Harvard Business Review, January 13, 2016.
https://hbr.org/2016/01/what-having-a-growth-mindset-actually-means.

Wikipedia. *"Ethics."*
https://en.wikipedia.org/wiki/Ethics."

Woebot Health. *"Products & Pipeline."*
https://woebothealth.com/products-pipeline/. Accessed November 12, 2021.

World Economic Forum. *"We Need a Global Reskilling Revolution—Here's Why."* January 22, 2020.
https://www.weforum.org/agenda/2020/01/reskilling-revolution-jobs-future-skills/.
Accessed November 12, 2021.

———. *"Robots Have Demonstrated Their Crucial Role in Pandemics—and How They Can Help for Years to Come."* May 6, 2020.
https://www.weforum.org/agenda/2020/05/robots-coronavirus-crisis/.
Accessed 12 Nov. 2021.

Zenbo Global. *"ASUS Zenbo—Your Smart Little Companion."*
https://zenbo.asus.com/.
Accessed November 12, 2021.

ZME Science. *"Decision Making Based on Instinct Is Surprisingly Accurate."* November 9, 2012.
https://www.zmescience.com/research/studies/decision-making-intuition-accurate-42433/.

Chapter 4:

Anderson, Charlie. *"The 85 Percent Rule of Worry."* LinkedIn, May 27, 2020.
https://www.linkedin.com/pulse/85-percent-rule-worry-charlie-anderson.

Clifford, Cat, and Zameena Mejia. *"Why Elon Musk Wants His Employees to Use a Strategy Called 'First Principles.'"* CNBC, April 18, 2018.
https://www.cnbc.com/2018/04/18/why-elon-musk-wants-his-employees-to-use-a-strategy-called-first-principles.html.

Epstein, David. R*ange: Why Generalists Triumph in a Specialized World.* Riverhead, 2019.

Gilbert, Dan. *"The Psychology of Your Future Self."* TED Talk, March 2014.
https://www.ted.com/talks/dan_gilbert_the_psychology_of_your_future_self/transcript?language=en.

Globokar, Lidija. *"The Power of Visualization and How to Use It."* Forbes, March 5, 2020.
https://www.forbes.com/sites/lidijaglobokar/2020/03/05/the-power-of-visualization-and-how-to-use-it/?sh=1ceb9cb26497.

"How to Worry Better." NBC, December 13, 2017.
https://www.nbcnews.com/better/pop-culture/praise-worry-why-fretting-can-be-good-you-ncna757016.

Interaction Design Foundation. *"Understanding Early Adopters and Customer Adoption Patterns."*
https://www.interaction-design.org/literature/article/understanding-early-adopters-and-customer-adoption-patterns. Accessed November 12, 2021.

Kahneman, Daniel. *Thinking, Fast and Slow.* Farrar, Straus and Giroux, 2013.

Klein, Gary. *The Power of Intuition: How to Use Your Gut Feelings to Make Better Decisions at Work.* DOUBLEDAY, a division of Random House, Inc., 2003.

Marr, Bernard. *Tech Trends in Practice: The 25 Technologies that are Driving the 4th Industrial Revolution.* John Wiley & Sons, 2020.

McKinsey & Company. *"Future Proof: Solving the 'Adaptability Paradox' for the Long Term."* August 2021.
https://www.mckinsey.com/~/media/mckinsey/business%20functions/organization/our%20insights/future%20proof%20solving%20the%20adaptability%20paradox%20for%20the%20long%20term/future-proof-solving-the-adaptability-paradox-for-the-long-term.pdf?shouldIndex=fa.

Olympic Channel. *"Visualisation—Olympic State of Mind."*
https://mind.olympicchannel.com/visualisation/. Accessed November 12, 2021.

Oshin, Mayo. *"Elon Musk's '3-Step' First Principles Thinking: How to Think and Solve Difficult Problems Like a Genius."* Medium, August 30, 2017.
https://medium.com/the-mission/elon-musks-3-step-first-principles-thinking-how-to-think-and-solve-difficult-problems-like-a-ba1e73a9f6c0.

Popomaronis, Tom. *"Elon Musk Calls This a 'Powerful, Powerful Way of Thinking'—But Is 'Hard to Do.' Here's How It Works."* CNBC, February 29, 2020.
https://www.cnbc.com/2020/02/28/billionaire-elon-musk-this-is-a-powerful-way-of-thinking-but-hard-to-do-how-it-works.html.

Reisinger, Don. "All Companies Should Live by the Jeff Bezos 70 Percent Rule." Inc.
https://www.inc.com/don-reisinger/all-companies-should-live-by-jeff-bezos-70-percent-rule.html. Accessed November 12, 2021.

School of Cooperative Individualism. *"Jacob Bronowski: The Reach of Imagination."*
https://www.cooperative-individualism.org/bronowski-jacob_reach-of-imagination.htm. Accessed November 12, 2021.

Schwantes, Marcel. *"Science Says Only 8 Percent of People Actually Achieve Their Goals. Here Are 7 Things They Do Differently."* Inc.
https://www.inc.com/marcel-schwantes/science-says-only-8-percent-of-people-actually-achieve-their-goals-here-are-7-things-they-do-differently.html.
Accessed November 12, 2021.

Singularity Hub. *"Ray Kurzweil's Mind-Boggling Predictions for the Next 25 Years."* January 26, 2015.
https://singularityhub.com/2015/01/26/ray-kurzweils-mind-boggling-predictions-for-the-next-25-years/.

Stanford University. *"Text of Steve Jobs' Commencement Address."* Stanford News, June 14, 2005.
https://news.stanford.edu/2005/06/14/jobs-061505/.

Subramanian, Courtney. *"Alvin Toffler: What He Got Right—and Wrong."* BBC, July 1, 2016.
https://www.bbc.com/news/world-us-canada-36675260.

Virgin. *"Great Ideas: Is Imagination More Important Knowledge?"*
https://www.virgin.com/about-virgin/latest/great-ideas-imagination-more-important-knowledge. Accessed November 12, 2021.

Chapter 5:

Accenture. *"Artificial Intelligence - AI ethics."*
https://www.accenture.com/us-en/insights/artificial-intelligence-summary-index?c=acn_glb_

Anderson, Mark Robert. *"After 75 Years, Isaac Asimov's Three Laws of Robotics Need Updating."* Conversation, March 17, 2017.
https://theconversation.com/after-75-years-isaac-asimovs-three-laws-of-robotics-need-updating-74501.

Booch, Grady. *"Don't Fear Superintelligent AI."* TED Talk, November 2016.
https://www.ted.com/talks/grady_booch_don_t_fear_superintelligent_ai/transcript? language=en.
Accessed November 12, 2021.

Friend, Tad. *"How Frightened Should We Be of A.I.?"* New Yorker, May 7, 2018.
https://www.newyorker.com/magazine/2018/05/14/how-frightened-should-we-be-of-ai.

Futurism. *"Kurzweil Claims That the Singularity Will Happen by 2045."*
https://futurism.com/kurzweil-claims-that-the-singularity-will-happen-by-2045.

———. *"Ray Kurzweil: 'AI Will Not Displace Humans, It's Going to Enhance Us.'"*
https://futurism.com/ray-kurzweil-ai-displace-humans-going-enhance.

———. *"Separating Science Fact from Science Hype: How Far off Is the Singularity?"*
January 30, 2018.
https://futurism.com/separating-science-fact-science-hype-how-far-off-singularity.

Ghosh, Dipayan. *"Are We Entering a New Era of Social Media Regulation?"* Harvard Business Review, January 14, 2021.
https://hbr.org/2021/01/are-we-entering-a-new-era-of-social-media-regulation.

Holmes, Aaron. *"Former Facebook Exec: We Made It as Addictive as Cigarettes on Purpose."* Business Insider, September 24, 2020.
https://www.businessinsider.com/former-facebook-exec-addictive-as-cigarettes-tim-kendall-2020-9?amp.

Investopedia. *"Turing Test."*
https://www.investopedia.com/terms/t/turing-test.asp.

Kasparov, Garry. *"Don't Fear Intelligent Machines. Work with Them."* TED Talk, April 2017.
https://www.ted.com/talks/garry_kasparov_don_t_fear_intelligent_machines_work_with_them/transcript.
Accessed November 12, 2021.

LaFrance, Adrienne. *"When People Feared Computers."* Atlantic, March 30, 2015.
https://www.theatlantic.com/technology/archive/2015/03/when-people-feared-computers/388919/.

Levy, Steven. *"Man vs. Machine."* Newsweek, May 4, 1997.
https://www.newsweek.com/man-vs-machine-173038.

Marr, Bernard. *"Is Artificial Intelligence (AI) a Threat to Humans?"* Forbes, March 2, 2020.

https://www.forbes.com/sites/bernardmarr/2020/03/02/is-artificial-intelligence-ai-a-threat-to-humans/?sh=589dd1f3205d.

Maryville University. *"The Evolution of Social Media: How Did It Begin and Where Could It Go Next?"* May 28, 2020.
https://online.maryville.edu/blog/evolution-social-media/.

Massachusetts Institute of Technology. *"Study: False News Spreads Faster Than the Truth."* Sloan School of Management, March 8, 2018.
https://mitsloan.mit.edu/ideas-made-to-matter/study-false-news-spreads-faster-truth.

Oberlo. *"How Many People Use Social Media in 2021."*
https://www.oberlo.com/statistics/how-many-people-use-social-media.
Accessed November 12, 2021.

———. *"10 Social Media Statistics You Need to Know in 2021."* April 6, 2021.
https://www.oberlo.com/blog/social-media-marketing-statistics#10-social-media-users-via-mobile.

PricewaterhouseCoopers. *"Sizing the Prize: What's the Real Value of AI for Your Business and How Can You Capitalise?"*
https://www.pwc.com/gx/en/issues/analytics/assets/pwc-ai-analysis-sizing-the-prize-report.pdf.

Ranker. *"9 Times in History When Everyone Freaked Out about New Technology."*
https://www.ranker.com/list/historical-freak-outs-about-new-technology/taeyura.
Accessed November 12, 2021.

Sacks, Rabbi Lord Jonathan. *"How We Can Face the Future without Fear, Together."* TED Talk, April 2017.
https://www.ted.com/talks/rabbi_lord_jonathan_sacks_how_we_can_face_the_future_without_fear_together/transcript#t-678162.
Accessed November 12, 2021.

Scraps from the Loft. *"The Social Dilemma (2020)—Transcript."* October 3, 2020.
https://scrapsfromtheloft.com/2020/10/03/the-social-dilemma-movie-transcript/.

Singularity Hub. *"Microsoft's Mesh Will Let Us Beam Ourselves to Work as Holograms."* March 10, 2021.
https://singularityhub.com/2021/03/10/microsoft-mesh-will-let-us-beam-ourselves-to-work-as-holograms/.

Statista. *"Value of Social Commerce Sales Worldwide from 2020 to 2026."* November 11, 2021.
https://www.statista.com/statistics/1251145/social-commerce-sales-worldwide/.

Vincent, James. *"Most Americans Think AI Will Destroy Other People's Jobs, Not Theirs."* Verge, March 7, 2018.
https://www.theverge.com/2018/3/7/17089904/ai-job-loss-automation-survey-gallup.

"What Do We Do about the Biases in AI?" Harvard Business Review, October 25, 2019.
https://hbr.org/2019/10/what-do-we-do-about-the-biases-in-ai.

Winder, Isabelle Catherine, and Vivien Shaw. *"Coronavirus: Experts in Evolution Explain Why Social Distancing Feels So Unnatural."* Conversation, March 25, 2020. https://theconversation.com/coronavirus-experts-in-evolution-explain-why-social-distancing-feels-so-unnatural-134271.

Chapter 6:

Bankrate. *"Survey: Fewer Than 4 in 10 Americans Could Pay a Surprise $1,000 Bill from Savings."* January 11, 2021. https://www.bankrate.com/banking/savings/financial-security-january-2021/.

Common Sense Media. *"New Survey Reveals Teens Get Their News from Social Media and YouTube."* August 12, 2019. https://www.commonsensemedia.org/about-us/news/press-releases/new-survey-reveals-teens-get-their-news-from-social-media-and-youtube.

Council for Economic Education. *"Survey of the States."* 2020. https://www.councilforeconed.org/wp-content/uploads/2020/02/2020-Survey-of-the-States.pdf.

Friedman, Zack. *"Student Loan Debt Statistics in 2019: A $1.5 Trillion Crisis."* Forbes, February 25, 2019. https://www.forbes.com/sites/zackfriedman/2019/02/25/student-loan-debt-statistics-2019/?sh=3dededfb133f.

Fundera. *"Crowdfunding Statistics (2021): Market Size and Growth."* July 26, 2019. https://www.fundera.com/resources/crowdfunding-statistics.

Global Footprint Network – Advancing the Science of Sustainability *"World Footprint."* https://www.footprintnetwork.org/our-work/ecological-footprint/. Accessed December 28, 2021.

Harari, Yuval Noah. *21 Lessons for the 21st Century*. Vintage Digital, 2018.

IEA. *"Net Zero by 2050."* May 2021. https://www.iea.org/reports/net-zero-by-2050.

Investopedia. *"Financial Literacy Definition."* February 12, 2009. https://www.investopedia.com/terms/f/financial-literacy.asp.

———. *"Initial Coin Offering (ICO) Definition."* December 20, 2016. https://www.investopedia.com/terms/i/initial-coin-offering-ico.asp.

John, Jennifer Neda. *"Opinion: Why Generation Z Falls for Online Misinformation."* MIT Technology Review, June 30, 2021. https://www.technologyreview.com/2021/06/30/1026338/gen-z-online-misinformation/.

LoveToKnow. *"The 1950s Family: Structure, Values and Everyday Life."* https://family.lovetoknow.com/about-family-values/1950s-family-structure-values-everyday-life. Accessed November 12, 2021.

McKinsey & Company. *"'True Gen': Generation Z and Its Implications for Companies."* November 12, 2018.
https://www.mckinsey.com/industries/consumer-packaged-goods/our-insights/true-gen-generation-z-and-its-implications-for-companies.

Pew Research Center. *"Information Overload."* December 7, 2016.
https://www.pewresearch.org/internet/2016/12/07/information-overload/.

———. *"Most US Teens Who Use Cellphones Do It to Pass Time, Connect with Others, Learn New Things."* August 23, 2019.
https://www.pewresearch.org/fact-tank/2019/08/23/most-u-s-teens-who-use-cellphones-do-it-to-pass-time-connect-with-others-learn-new-things/.

———. *"On the Cusp of Adulthood and Facing an Uncertain Future: What We Know about Gen Z So Far."* May 14, 2020.
https://www.pewresearch.org/social-trends/2020/05/14/on-the-cusp-of-adulthood-and-facing-an-uncertain-future-what-we-know-about-gen-z-so-far-2/.

———. *"Social Media Use in 2021."* April 7, 2021.
https://www.pewresearch.org/internet/2021/04/07/social-media-use-in-2021/.

Reinsel, David, John Gantz, and John Rydning. *"The Digitization of the World: From Edge to Core."* White paper, IDC, November 2018.
https://www.seagate.com/files/www-content/our-story/trends/files/idc-seagate-dataage-whitepaper.pdf.

Stanford University. *"Stanford Researchers Find Students Have Trouble Judging the Credibility of Information Online."* November 22, 2016.
https://ed.stanford.edu/news/stanford-researchers-find-students-have-trouble-judging-credibility-information-online.

White, Alexandria. *"73% of Americans Rank Their Finances As the No. 1 Stress in Life, According to New Capital One CreditWise Survey."* CNBC, July 20, 2021.
https://www.cnbc.com/select/73-percent-of-americans-rank-finances-as-the-number-one-stress-in-life/.

Whitebread, David and Bingham, Sue. *"Habit Formation and Learning in Young Children."*
https://mascdn.azureedge.net/cms/the-money-advice-service-habit-formation-and-learning-in-young-children-may2013.pdf

World Economic Forum. *"How Much Data Is Generated Each Day?"* April 17, 2019.
https://www.weforum.org/agenda/2019/04/how-much-data-is-generated-each-day-cf4bddf29f/.
Accessed November 12, 2021.

———. *"Jobs Will Be Very Different in 10 Years. Here's How to Prepare."* January 17, 2020.
https://www.weforum.org/agenda/2020/01/future-of-work/.

———. *"We Asked Young People What Changes They Want for the Future. Here's What They Said."* August 21, 2021.

https://www.weforum.org/agenda/2021/08/young-people-millennials-and-generation-z-change-future/?utm_source=sfmc&utm_medium=email&utm_campaign=2753690_Agenda_weekly-27August2021&utm_term=&emailType=Agenda%20Weekly.

Wunderman Thompson Intelligence *"Into the Metaverse."* September 14, 2021. https://www.wundermanthompson.com/insight/new-trend-report-into-the-metaverse. Accessed December 28, 2021.

Chapter 7:

"9 Out of 10 People Are Willing to Earn Less Money to Do More-Meaningful Work." Harvard Business Review, November 6, 2018. https://hbr.org/2018/11/9-out-of-10-people-are-willing-to-earn-less-money-to-do-more-meaningful work.

Adam, David. *"What If Aging Weren't Inevitable, but a Curable Disease?"* MIT Technology Review, August 19, 2019. https://www.technologyreview.com/2019/08/19/133357/what-if-aging-werent-inevitable-but-a-curable-disease/.

American Psychological Association. *"Julianne Holt-Lunstad Probes Loneliness, Social Connections."* October 18, 2017. https://www.apa.org/members/content/holt-lunstad-loneliness-social-connections.

Association of American Medical Colleges. *"US Physician Shortage Growing."* June 26, 2020. https://www.aamc.org/news-insights/us-physician-shortage-growing.

Balakrishnan, Anita. *"Silicon Valley Bigwigs Like Larry Ellison, Peter Thiel, and Sergey Brin Are Trying to End Aging."* CNBC, March 27, 2017. https://www.cnbc.com/2017/03/27/anti-aging-technology-larry-ellison-peter-thiel-sergey-brin.html.

Blue Zones. *"Reverse Engineering Longevity."* November 10, 2016. https://www.bluezones.com/2016/11/power-9/.

Buettner, Dan. *"How to Live to Be 100+."* TED Talk, September 2009. https://www.ted.com/talks/dan_buettner_how_to_live_to_be_100/transcript?referrer=playlist-what_s_the_secret_to_living_longer&language=en#t-806291. Accessed November 12, 2021.

Buettner, Dan, and Sam Skemp. *"Blue Zones: Lessons from the World's Longest Lived."* American Journal of Lifestyle Medicine 10, no. 5 (2016): 318–21, doi:10.1177/1559827616637066. https://www.ncbi.nlm.nih.gov/pmc/articles/PMC6125071/.

Calico. *"Drug Development."* https://www.calicolabs.com/drug-development. Accessed November 12, 2021.

Centers for Disease Control and Prevention. *"Loneliness and Social Isolation Linked to Serious Health Conditions."*

https://www.cdc.gov/aging/publications/features/lonely-older-adults.html. Accessed November 12, 2021.

Cigna. *"Loneliness and the Workplace: 2020 U.S. Report."* 2020. https://www.cigna.com/static/www-cigna-com/docs/about-us/newsroom/studies-and-reports/combatting-loneliness/cigna-2020-loneliness-report.pdf

Deloitte. *"The Social Enterprise in a World Disrupted."* 2021. https://www2.deloitte.com/content/dam/Deloitte/lu/Documents/human-capital/lu-hc-trends-2021.pdf.

Diamandis, Peter H. *"How long will you live? - The Longevity Mindset."* Blog post. https://www.diamandis.com/blog/lifespan-longevity-mindset. Accessed November 12, 2021.

———. *"Human Longevity Inc."* https://www.diamandis.com/human-longevity-inc.

Diamandis, Peter H., and Steven Kotler. *The Future Is Faster Than You Think.* Simon & Schuster, 2020.

Gannon, Megan. *"China's First Emperor Ordered Official Search for Immortality Elixir."* Live Science, December 27, 2017. https://www.livescience.com/61286-first-chinese-emperor-sought-immortality.html.

Getlen, Larry. *"Scientists Could One Day Make Humans Immortal."* New York Post, January 6, 2018. https://nypost.com/2018/01/06/scientists-could-one-day-make-humans-immortal/.

Harari, Yuval Noah. *Homo Deus: A Brief History of Tomorrow.* Harper, 2017.

Humphries, Courtney. *"Digital Immortality: How Your Life's Data Means a Version of You Could Live Forever."* MIT Technology Review, October 18, 2018. https://www.technologyreview.com/2018/10/18/139457/digital-version-after-death/.

International Monetary Fund. *"The Financial Sector in the 2020s: Building a More Inclusive System in the New Decade."* January 17, 2020. https://www.imf.org/en/News/Articles/2020/01/17/sp01172019-the-financial-sector-in-the-2020s.

Isaak, Adam. *"The Ultra-Rich Are Investing in Companies Trying to Reverse Aging. Is It Going to Work?"* CNBC, February 23, 2020. https://www.cnbc.com/2020/02/19/the-ultra-rich-are-investing-in-companies-trying-to-reverse-aging.html.

Kurzweil. *"Apparently We're All Going to Live Forever by 2029."* https://www.kurzweilai.net/metro-apparently-were-all-going-to-live-forever-by-2029. Accessed November 12, 2021.

———. *"We'll Be Uploading Our Entire Minds to Computers by 2045 and Our Bodies Will Be Replaced by Machines within 90 Years, Google Expert Claims."*

https://www.kurzweilai.net/daily-mail-well-be-uploading-our-entire-minds-to-computers-by-2045-and-our-bodies-will-be-replaced-by-machines-within-90-years-google-expert-claims. Accessed November 12, 2021.

Leonhard, Gerd. Technology vs. *Humanity: The Coming Clash between Man and Machine.* Fast Future, 2016.

Livni, Ephrat. *"Humans Don't Actually Want to Be Immortal, We Just Want to Be Forever Young."* Quartz, June 27, 2019.
https://qz.com/1640810/we-want-eternal-youth-not-immortality/.

Maris, Bill. *"Google Ventures and the Search for Immortality."* Bloomberg, March 9, 2015.
https://www.bloomberg.com/news/articles/2015-03-09/google-ventures-bill-maris-investing-in-idea-of-living-to-500.

Marr, Bernard. *Tech Trends in Practice: The 25 Technologies that are Driving the 4th Industrial Revolution.* John Wiley & Sons, 2020.

National Human Genome Research Institute. *"The Cost of Sequencing a Human Genome."*
https://www.genome.gov/about-genomics/fact-sheets/Sequencing-Human-Genome-cost. Accessed November 12, 2021.

National Intelligence Council. *"Global Trends 2040."* March 2021.
https://www.dni.gov/files/ODNI/documents/assessments/GlobalTrends_2040.pdf.

"Opening the Door to Treating Ageing As a Disease." Editorial, Lancet 6, no. 8 (2018): 587.
https://www.thelancet.com/journals/landia/article/PIIS2213-8587(18)30214-6/fulltext.

Our World in Data. *"Life Expectancy at Birth, Including the UN Projections."*
https://ourworldindata.org/grapher/life-expectancy-at-birth-including-the-un-projections. Accessed November 12, 2021.

———. *"Life Expectancy."*
https://ourworldindata.org/life-expectancy. October 2019.
Accessed November 12, 2021.

Pinker, Susan. *"The Secret to Living Longer May Be Your Social Life."* TED Talk, April 2017.
https://www.ted.com/talks/susan_pinker_the_secret_to_living_longer_may_be_your_social_life?referrer=playlist-what_s_the_secret_to_living_longer#t-883923.
Accessed November 12, 2021.

Regalado, Antonio. "*Google's Long, Strange Life-Span Trip.*" MIT Technology Review, December 15, 2016.
https://www.technologyreview.com/2016/12/15/69305/googles-long-strange-life-span-trip/.

Sinclair, David A., and Nir Barzilai. "This Is Why We Need to Start Treating Ageing As a Disease." Wired, December 31, 2018.
https://www.wired.co.uk/article/medicine-disease-ageing.

Singh-Kurtz, Sangeeta. *"Why Silicon Valley Titans Are Obsessed with Immortality."* Quartz.

https://qz.com/1578796/why-silicon-valley-titans-are-obsessed-with-immortality/. Accessed November 12, 2021.

Singularity Hub. *"How Long Can We Live? New Research Says the Human Lifespan Tops Out at 150."* June 7, 2021. https://singularityhub.com/2021/06/07/how-long-can-humans-live-new-research-says-our-lifespan-tops-out-at-150/.

Sociable. *"'The Future's Gonna Be Weird': Musk Says Memories Could Be Downloaded into New Body or Robot."* September 4, 2020. https://sociable.co/technology/futures-gonna-be-weird-musk-memories-downloaded-new-body-robot/.

Statista. *"Is 100 the New 80? Centenarians Are Becoming More Common."* https://www.statista.com/chart/18826/number-of-hundred-year-olds-centenarians-worldwide/. Accessed November 12, 2021.

———. *"Projected Global Life Expectancy."* https://www.statista.com/statistics/673420/projected-global-life-expectancy/. Accessed November 12, 2021.

———. *"Where 100 Is the New 80."* https://www.statista.com/chart/14931/where-100-is-the-new-80/. Accessed November 12, 2021.

United Nations. *"Mental Health and Development."* https://www.un.org/development/desa/disabilities/issues/mental-health-and-development.html.

———. *"Growing at a Slower Pace, World Population Is Expected to Reach 9.7 Billion in 2050 and Could Peak at Nearly 11 Billion around 2100."* June 17, 2019. https://www.un.org/development/desa/en/news/population/world-population-prospects-2019.html. Accessed November 12, 2021.

United States Food and Drug Administration. *"FDA Approves First Treatment for Hutchinson-Gilford Progeria Syndrome and Some Progeroid Laminopathies."* https://www.fda.gov/news-events/press-announcements/fda-approves-first-treatment-hutchinson-gilford-progeria-syndrome-and-some-progeroid-laminopathies. Accessed November 12, 2021.

United States Health Resources and Service Administration. *"Organ Donation Statistics."* https://www.organdonor.gov/statistics-stories/statistics.html. Accessed November 12, 2021.

United States Social Security Administration. *"Trustees Report Summary."* https://www.ssa.gov/OACT/TRSUM/2019/index.html. Accessed November 12, 2021.

Unity Biotechnology. *"Targeting a Root Cause."* https://unitybiotechnology.com/. Accessed November 12, 2021.

University of California, Los Angeles. *"Martine Rothblatt Works to Make the World a Better Place."* March 24, 2021.
https://newsroom.ucla.edu/magazine/martine-rothblatt-transhumanism.

Waldinger, Robert. *"What Makes a Good Life? Lessons from the Longest Study on Happiness."* TED Talk, November 2015.
https://www.ted.com/talks/robert_waldinger_what_makes_a_good_life_lessons_from_the_longest_study_on_happiness/transcript?referrer=playlist-what_s_the_secret_to_living_longer&language=en#t-750940.
Accessed November 12, 2021.

Wikipedia. *"Fountain of Youth."*
https://en.wikipedia.org/wiki/Fountain_of_Youth.

World Economic Forum. *"How 3D-printed organs could save the lives of transplant patients."*
https://www.weforum.org/agenda/2021/04/medical-breakthrough-how-tissue-engineering-could-help-the-world-s-organ-shortage/.

Your Genome. *"What Is CRISPR-Cas9?"*
https://www.yourgenome.org/facts/what-is-crispr-cas9.

Chapter 8:

Accenture. *"Shaping the Sustainable Organization."*
https://www.accenture.com/_acnmedia/Thought-Leadership-Assets/PDF-5/Accenture-Shaping-the-Sustainable-Organization-Report.pdf.

Baicker, Katherine, Cutler, David, and Song, Zirui. *"Health Affairs. Workplace Wellness Programs Can Generate Savings."*
https://www.healthaffairs.org/doi/10.1377/hlthaff.2009.0626.
Accessed December 29, 2021.

Bates Ramirez, Vanessa. *"The 6 Ds of Tech Disruption: A Guide to the Digital Economy."*
https://singularityhub.com/2016/11/22/the-6-ds-of-tech-disruption-a-guide-to-the-digital-economy/.

BCG. *"The End of Management As We Know It."*
https://www.bcg.com/publications/2020/end-management-as-we-know-it.

Carlton, Bobby. *"Virtual Reality and Learning – A Masie Report."* January 2017.
http://ivrtrain.com/ivrtrain/VR_Learn_Report-2017.pdf

Chernev, Bobby. *"29 Astonishing E-learning Statistics for 2022."* February 06, 2022. https://techjury.net/blog/elearning-statistics/#gref.

Cognizant. *"Relearning How We Learn, From the Campus to the Workplace."*
https://www.cognizant.com/whitepapers/relearning-how-we-learn-from-the-campus-to-the-workplace-codex3921.pdf.

Deloitte. *"Automation with Intelligence."* November 25, 2020.
https://www2.deloitte.com/us/en/insights/focus/technology-and-the-future-of-work/intelligent-automation-2020-survey-results.html.

Education Data Initiative. *"College Dropout Rate [2021]: By Year + Demographics."* November 6, 2019.
https://educationdata.org/college-dropout-rates.

Epstein, David. *Range: Why Generalists Triumph in a Specialized World.* Riverhead, 2019.

Evans, Dain. *"Apple Could Win Big in Health Care."* CNBC, October 13, 2020.
https://www.cnbc.com/2020/10/13/apple-could-win-big-in-health-care.html.

Forrester. *"The Future of Work".*
https://www.forrester.com/technology/future-of-work/

Freeze, Di. *"Cybercrime to Cost the World $10.5 Trillion Annually by 2025."* Cybercrime, December 14, 2020.
https://cybersecurityventures.com/cybercrime-will-cost-the-world-16-4-billion-a-day-in-2021/.

Friedman, Zack. *"Student Loan Debt Statistics in 2021: A Record $1.7 Trillion."* Forbes, February 20, 2021.
https://www.forbes.com/sites/zackfriedman/2021/02/20/student-loan-debt-statistics-in-2021-a-record-17-trillion/?sh=7ee3fa7a1431.

Gartner. *"Routine Work Done by Managers to Be Fully Automated."* Press release, January 23, 2020.
https://www.gartner.com/en/newsroom/press-releases/2020-01-23-gartner-predicts-69--of-routine-work-currently-done-b. Accessed November 12, 2021.

Gov.UK. *"Current and Future Demand for Digital Skills—A Call to Action."* June 13, 2019.
https://www.gov.uk/government/publications/current-and-future-demand-for-digital-skills-in-the-workplace.

Harari, Yuval Noah. *Sapiens: A Brief History of Humankind.* HarperCollins, 2015.

Harari, Yuval Noah. *21 Lessons for the 21st Century.* Random House, 2019.

Hess, Abigail Johnson. *"Cost of College Increased by More than 25% in the Last 10 Years."* CNBC, December 13, 2019.
https://www.cnbc.com/2019/12/13/cost-of-college-increased-by-more-than-25percent-in-the-last-10-years.html.

History.com. *"Thomas Edison."* June 6, 2019.
https://www.history.com/topics/inventions/thomas-edison.

IEA. *"Net Zero by 2050—Analysis."*
https://www.iea.org/reports/net-zero-by-2050. Accessed November 12, 2021.

IBM. *"The Enterprise Guide to Closing the Skills Gap."*
https://www.ibm.com/downloads/cas/EPYMNBJA.

IRENA. *"Renewable Energy and Jobs—Annual Review 2021."* https://irena.org/publications/2021/Oct/Renewable-Energy-and-Jobs-Annual-Review-2021.

International Labour Organization. *"24 million jobs to open up in the green economy."* https://www.ilo.org/global/about-the-ilo/newsroom/news/WCMS_628644/lang--en/index.htm. Accessed November 12, 2021.

Learning & Work Institute. *"Disconnected? Exploring the digital skills gap."* https://learningandwork.org.uk/resources/research-and-reports/disconnected-exploring-the-digital-skills-gap/

Liu, Jennifer. *"Gallup: U.S. Workers Are among the Most Stressed in the World."* CNBC, June 15, 2021. https://www.cnbc.com/amp/2021/06/15/gallup-us-workers-are-among-the-most-stressed-in-the-world.html.

Marr, Bernard. *Tech Trends in Practice: The 25 Technologies that are Driving the 4th Industrial Revolution.* John Wiley & Sons, 2020.

McKinsey & Company. *"Beyond hiring: How companies are reskilling to address talent gaps."* February 12, 2020. https://www.mckinsey.com/business-functions/people-and-organizational-performance/our-insights/beyond-hiring-how-companies-are-reskilling-to-address-talent-gaps.

———. *"Competing in a World of Sectors without Borders."* July 12, 2017. https://www.mckinsey.com/business-functions/mckinsey-analytics/our-insights/competing-in-a-world-of-sectors-without-borders.

———. *"Artificial Intelligence in Education: How Will It Impact K–12 Teachers."* January 14, 2020. https://www.mckinsey.com/industries/public-and-social-sector/our-insights/how-artificial-intelligence-will-impact-k-12-teachers.

———. *"Skill shift: Automation and the future of the workforce."* May 23, 2018. https://www.mckinsey.com/~/media/mckinsey/industries/public%20and%20social%20sector/our%20insights/skill%20shift%20automation%20and%20the%20future%20of%20the%20workforce/mgi-skill-shift-automation-and-future-of-the-workforce-in-brief-may-2018.pdf.

———. *"The Future of Remote Work: An Analysis of 2,000 Tasks, 800 Jobs, and 9 Countries."* November 23, 2020. https://www.mckinsey.com/featured-insights/future-of-work/whats-next-for-remote-work-an-analysis-of-2000-tasks-800-jobs-and-nine-countries.

Morgan Stanley. *"Flying Car and Autonomous Aircraft Research."* https://www.morganstanley.com/ideas/autonomous-aircraft. Accessed November 12, 2021.

National Skills Coalition. *"The New Landscape of Digital Literacy."* May 2020. https://nationalskillscoalition.org/wp-content/uploads/2020/12/05-20-2020-NSC-New-Landscape-of-Digital-Literacy.pdf.

Net Zero Tracker. Home page.
https://zerotracker.net/.
Accessed November 24, 2021.

PricewaterhouseCoopers. *"2021 CEO Survey: US Findings."*
https://www.pwc.com/us/en/library/ceo-agenda/ceo-survey.html.
Accessed November 12, 2021.

———. *"How Will Automation Impact Jobs."*
https://www.pwc.co.uk/services/economics/insights/the-impact-of-automation-on-jobs.html.
Accessed November 12, 2021.

———. *"Will robots really steal our jobs?"*
https://www.pwc.com/hu/hu/kiadvanyok/assets/pdf/impact_of_automation_on_jobs.pdf.
Accessed February 13, 2022.

Reinicke, Carmen. *"Autonomous Vehicles Won't Only Kill Jobs. They Will Create Them Too."* CNBC, August 11, 2018.
https://www.cnbc.com/2018/08/10/autonomous-vehicles-are-creating-jobs-heres-where.html.

Talent Management. *"Skills Aren't Hard or Soft—They're Durable or Perishable."* https://www.talentmgt.com/articles/2020/10/29/skills-arent-soft-or-hard-theyre-durable-or-perishable/.

Tencent. *"Connecting Ecosystems."*
https://www.tencent.com/en-us/business.html?page-active=to-b.

UNESCO. *"Digital Skills Critical for Jobs and Social Inclusion."* March 15, 2018.
https://en.unesco.org/news/digital-skills-critical-jobs-and-social-inclusion.

United States Bureau of Labor Statistics. *"Why Computer Occupations Are behind Strong STEM Employment Growth in the 2019–29 Decade."* January 19, 2021.
https://www.bls.gov/opub/btn/volume-10/why-computer-occupations-are-behind-strong-stem-employment-growth.htm.

———. *"Employment to grow 7.7 percent from 2020 to 2030; 1.7 percent excluding COVID-19 recovery."* September 23, 2021.
https://www.bls.gov/opub/ted/2021/employment-to-grow-7-7-percent-from-2020-to-2030-1-7-percent-excluding-covid-19-recovery.htm.

———. *"Employment in STEM Occupations."*
https://www.bls.gov/emp/tables/stem-employment.htm.

———. *"Healthcare Occupations : Occupational Outlook Handbook."* September 8, 2021.
https://www.bls.gov/ooh/healthcare/home.htm.

———. *"Information Security Analysts: Occupational Outlook Handbook."* September 8, 2021.
https://www.bls.gov/ooh/computer-and-information-technology/information-security-analysts.htm.

World Economic Forum. *"Shaping the Future of Digital Economy and New Value Creation."*
https://www.weforum.org/platforms/shaping-the-future-of-digital-economy-and-new-value-creation.

———. *"Closing the Skills Gap Accelerators."*
https://www.weforum.org/projects/closing-the-skills-gap-accelerators.
Accessed February 11, 2022.

———. *"This Was the Year Automation Started Taking Over the Workforce."* May 26, 2020.
https://www.weforum.org/agenda/2020/05/automation-robot-employment-inequality.
Accessed November 12, 2021.

———. *"The Future of Jobs Report 2020."* October 20, 2020.
https://www.weforum.org/reports/the-future-of-jobs-report-2020.
Accessed November 12, 2021.

———. *"Recession and Automation Changes Our Future of Work, but There Are Jobs Coming, Report Says."* October 20, 2020.
https://www.weforum.org/press/2020/10/recession-and-automation-changes-our-future-of-work-but-there-are-jobs-coming-report-says-52c5162fce/.
Accessed November 12, 2021.

———. *"40% of All Projected Job Opportunities Will Be Created in This Sector. And It's Not Technology."* May 28, 2021.
https://www.weforum.org/agenda/2021/05/care-economy-emerging-job-opportunities/?utm_source=sfmc&utm_medium=email&utm_campaign=2747176_Agenda_weekly-4June2021&utm_term=&emailType=Newsletter.
Accessed November 12, 2021.

———. *"Modern Farming Is As Much about Data As Digging. Here Are 3 Emerging Agricultural Skills."* June 2, 2021.
https://www.weforum.org/agenda/2021/06/farming-data-new-agricultural-job-skills/.
Accessed November 12, 2021.

———. *"These Are the Fastest Growing Jobs in the United States."* September 20, 2021.
https://www.weforum.org/agenda/2021/09/fastest-growing-jobs-employment-skills/?utm_source=sfmc&utm_medium=email&utm_campaign=2755935_Agenda_weekly-24September2021&utm_term=&emailType=Agenda%20Weekly.

Zimmerman, Eli. *"The 4 C's of Learning in a Connected Classroom."* EdTech, July 27, 2018.
https://edtechmagazine.com/k12/article/2018/07/4-cs-learning-connected-classroom.
Accessed November 12, 2021.

Chapter 9:

Deloitte. *"The State of Women in AI Today."*
https://www2.deloitte.com/us/en/pages/consulting/articles/state-of-women-in-ai-today.html.

Dell. *"Dell Technologies' Commitment to Customers' Privacy."*
https://www.dell.com/en-us/shop/data-privacy/cp/data-privacy.

Dizikes, Peter. *"Study finds stronger links between automation and inequality."* May 5, 2020.
https://news.mit.edu/2020/study-inks-automation-inequality-0506

Federal Reserve Board. *"Disparities in Wealth by Race and Ethnicity in the 2019 Survey of Consumer Finances."*
https://www.federalreserve.gov/econres/notes/feds-notes/disparities-in-wealth-by-race-and-ethnicity-in-the-2019-survey-of-consumer-finances-20200928.htm.
Accessed November 12, 2021.

———. *"Distribution of Household Wealth in the US since 1989."*
https://www.federalreserve.gov/releases/z1/dataviz/dfa/distributechart/#quarter:122;series:Net%20worth;demographic:race;population:all;units:shares;range:1989.3,2020.1.
Accessed November 12, 2021.

Google. *"Artificial Intelligence at Google: Our Principles."*
https://ai.google/principles/. Accessed November 12, 2021.

IBM. *"AI Ethics."*
https://www.ibm.com/artificial-intelligence/ethics.
Accessed November 12, 2021.

International Labour Organization. *"The Gender Gap in Employment: What's Holding Women Back?"*
https://www.ilo.org/infostories/en-GB/Stories/Employment/barriers-women#global-gap/labour-force.
Accessed November 12, 2021.

McKinsey & Company. *"The Economic Impact of Closing the Racial Wealth Gap."* August 13, 2019.
https://www.mckinsey.com/industries/public-and-social-sector/our-insights/the-economic-impact-of-closing-the-racial-wealth-gap.

Microsoft. *"Putting principles into practice: How we approach responsible AI at Microsoft."*
https://query.prod.cms.rt.microsoft.com/cms/api/am/binary/RE4pKH5.
Accessed November 12, 2021.

National Bureau of Economic Research. *"Tasks, Automation, and the Rise in US Wage Inequality."*
https://www.nber.org/papers/w28920.
Accessed November 12, 2021.

National Intelligence Council. *"Global Trends 2040."* March 2021.
https://www.dni.gov/index.php/gt2040-home/gt2040-structural-forces/demographics-and-human-development.

Saraiva, Catarina. *"Women Could Give $20 Trillion Boost to Economic Growth by 2050."* Bloomberg, March 9, 2021.
https://www.bloomberg.com/news/articles/2021-03-09/women-could-give-20-trillion-boost-to-economic-growth-by-2050.
Accessed November 12, 2021.

UNESCO. *"Artificial intelligence and gender equality: key findings of UNESCO's Global Dialogue."*
https://unesdoc.unesco.org/ark:/48223/pf0000374174.

UNICEF. *"Two Thirds of the World's School-Age Children Have No Internet Access at Home, New UNICEF-ITU Report Says."* November 30, 2020.

https://www.unicef.org/press-releases/two-thirds-worlds-school-age-children-have-no-internet-access-home-new-unicef-itu.

United Nations. *"With Almost Half of World's Population Still Offline, Digital Divide Risks Becoming 'New Face of Inequality,' Deputy Secretary-General Warns General Assembly."* Press release, April 27, 2021.
https://www.un.org/press/en/2021/dsgsm1579.doc.htm.

———. *"World Social Report 2020: Inequality in a Rapidly Changing World."*
https://www.un.org/development/desa/dspd/wp-content/uploads/sites/22/2020/02/World-Social-Report2020-FullReport.pdf.

United States Census Bureau. *"Women Making Gains in STEM Occupations but Still Underrepresented."* January 26, 2021.
https://www.census.gov/library/stories/2021/01/women-making-gains-in-stem-occupations-but-still-underrepresented.html.

Vartan, Starre. *"Racial Bias Found in a Major Health Care Risk Algorithm."* Scientific American, October 24, 2019.
https://www.scientificamerican.com/article/racial-bias-found-in-a-major-health-care-risk-algorithm/. Accessed November 12, 2021.

White House. *"How the Build Back Better Framework Will Support Women's Employment and Strengthen Family Economic Security."* Fact sheet, July 15, 2021. https://www.whitehouse.gov/briefing-room/statements-releases/2021/07/15/fact-sheet-how-the-build-back-better-framework-will-support-womens-employment-and-strengthen-family-economic-security/.

World Economic Forum. *"Global Gender Gap Report 2021."*
https://www3.weforum.org/docs/WEF_GGGR_2021.pdf.

Chapter 10:

Blue Origin. *"For The Benefit Of Earth."*
https://www.blueorigin.com/about-blue. Accessed February 11, 2022.

Canadian Space Agency. *"Everyday Benefits of Space Exploration."*
https://www.asc-csa.gc.ca/eng/about/everyday-benefits-of-space-exploration/improving-health-care.asp.

Drake, Nadia. *"SpaceX Launches New Era of Spaceflight with Company's First Crewed Mission."* National Geographic, May 30, 2020.
https://www.nationalgeographic.com/science/article/spacex-nasa-launch-human-astronauts-crew-dragon-international-space-station-demo-2.

———. *"Why we explore Mars—and what decades of missions have revealed."* National Geographic.
https://www.nationalgeographic.com/science/article/mars-exploration-article.

Einhorn, Bruce. *"Elon Musk's Mars Ambition Could Be the Riskiest Human Quest Ever."* Bloomberg, May 25, 2021.

https://www.bloomberg.com/news/articles/2021-05-25/elon-musk-s-mars-ambition-could-be-the-riskiest-human-quest-ever. Accessed November 12, 2021.

Jones, Harry W. *"The Recent Large Reduction in Space Launch Cost."* 48th International Conference on Environmental Systems, July 8–12, 2018, Albuquerque, New Mexico. https://ttu-ir.tdl.org/handle/2346/74082.

Khan Academy. *"Homo Sapiens and Early Human Migration."*
https://www.khanacademy.org/humanities/world-history/world-history-beginnings/origin-humans-early-societies/a/where-did-humans-come-from.

Kharpal, Arjun. *"China plans to send its first crewed mission to Mars in 2033 and build a base there."* CNBC, June 24, 2021.
https://www.cnbc.com/2021/06/24/china-plans-to-send-its-first-crewed-mission-to-mars-in-2033.html.

Morgan Stanley. *"Investing in Space Exploration."*
https://www.morganstanley.com/ideas/investing-in-space.
Accessed November 12, 2021.

———. *"Space Sustainability and Exploration."*
https://www.morganstanley.com/ideas/space-earth-sustainability.
Accessed November 12, 2021.

Murphy, Dan and Turak, Natasha. *"United Arab Emirates becomes the first Arab country to reach Mars."* CNBC, February 9, 2021.
https://www.cnbc.com/2021/02/09/mars-probe-uae-attempts-to-become-first-arab-country-to-reach-mars-with-hope-probe.html.

National Aeronautics and Space Administration. *"Moon to Mars Overview."*
https://www.nasa.gov/topics/moon-to-mars/overview. Accessed November 12, 2021.

———. *"What Is Mars?"*
https://www.nasa.gov/audience/forstudents/5-8/features/nasa-knows/what-is-mars-58.html.
Accessed November 12, 2021.

———. *"Benefits Stemming from Space Exploration."* International Space Exploration Coordination Group, September 2013.
https://www.nasa.gov/sites/default/files/files/Benefits-Stemming-from-Space-Exploration-2013-TAGGED.pdf.

———. *"As Artemis Moves Forward, NASA Picks SpaceX to Land Next Americans on Moon."* Press release, April 16, 2021.
https://www.nasa.gov/press-release/as-artemis-moves-forward-nasa-picks-spacex-to-land-next-americans-on-moon.
Accessed November 12, 2021.

———. *"NASA Publishes Artemis Plan to Land First Woman, Next Man on Moon in 2024."* Press release, September 21, 2020.
https://www.nasa.gov/press-release/nasa-publishes-artemis-plan-to-land-first-woman-next-man-on-moon-in-2024.

Accessed November 12, 2021.

Patel, Neel V. *"The Best Places to Find Extraterrestrial Life in Our Solar System, Ranked."* MIT Technology Review, June 16, 2021. https://www.technologyreview.com/2021/06/16/1026473/best-worlds-extraterrestrial-life-solar-system-ranked/.

Rincon, Paul. *"Jeff Bezos Launches to Space Aboard New Shepard Rocket Ship."* BBC, July 20, 2021. https://www.bbc.com/news/science-environment-57849364.

Sheetz, Michael. *"Space Tourism: How SpaceX, Virgin Galactic, Blue Origin, Axiom Compete."* CNBC, September 26, 2020. https://www.cnbc.com/2020/09/26/space-tourism-how-spacex-virgin-galactic-blue-origin-axiom-compete.html.

———. *"Elon Musk Is 'Highly Confident' SpaceX Will Land Humans on Mars by 2026."* CNBC, December 1, 2020. https://www.cnbc.com/2020/12/01/elon-musk-highly-confident-spacex-will-land-humans-on-mars-by-2026.html.

———. *"Elon Musk Wants SpaceX to Reach Mars So Humanity Is Not a 'Single-Planet Species.'"* CNBC, April 23, 2021. https://www.cnbc.com/2021/04/23/elon-musk-aiming-for-mars-so-humanity-is-not-a-single-planet-species.html.

Stankiewicz, Kevin. *"NASA Administrator Bill Nelson: Richard Branson's Space Flight a Great Milestone."* CNBC, July 12, 2021. https://www.cnbc.com/2021/07/12/nasa-administrator-bill-nelson-richard-branson-space-flight-a-great-milestone.html.

UNOOSA. *"Outer Space Treaty."* http://www.unoosa.org/oosa/en/ourwork/spacelaw/treaties/outerspacetreaty.html. Accessed November 12, 2021.

Wikipedia. *"Outer Space Treaty."* https://en.wikipedia.org/wiki/Outer_Space_Treaty.

Williams, Matthew S. *"Making Green on the Red Planet: How Might We Build an Economy on Mars?"* Interesting Engineering, December 2, 2019. https://interestingengineering.com/making-green-on-the-red-planet-how-might-we-build-an-economy-on-mars.

Epilogue:

Diamandis, Peter H., and Steven Kotler. *The Future Is Faster Than You Think.* Simon & Schuster, 2020.

Made in the USA
Monee, IL
12 November 2026

f7197256-8644-4e95-b5d6-f65778900623R01